Lecture Notes in Economics and Mathematical Systems

392

Karl Frauendorfer

Stochastic Two-Stage Programming

Springer-Verlag
Berlin Heidelberg New York
London Paris Tokyo
Hong Kong Barcelona
Budapest

Author

Dr. Karl Frauendorfer
Institute for Operations Research
University of Zurich
Moussonstr. 15
Ch-8044 Zurich

ISBN-13: 978-3-540-56097-5 e-ISBN-13: 978-3-642-95696-6
DOI: 10.1007/978-3-642-95696-6

Typesetting: Camera ready by author/editor
42/3140-543210 - Printed on acid-free paper

PREFACE

Analysts of systems in economics, ecology, industry etc. are confronted with increasing complexity due to the progress achieved in these fields. For finding (almost) optimal decisions, analysts are faced firstly with the problem statement, secondly with the set-up of a model, and thirdly with the insertion of proper and efficient solution methods.

Stochastic Programming (SP) offers models and methods that allow for application in decision problems where some of the problem data is uncertain, and known only in a probabilistic sense. SP models can be classified into *single-period models, two-stage models, models with probabilistic constraints* and *multi-period models.* These models are endowed with advantageous features and structural properties preferably exploited by SP methods within the solution process.

This work contributes to the methodology for two-stage models within Stochastic Programming. In these models the objective function is given as an integral whose integrand depends on a random vector, on its probability measure and on a decision. The fact that the integrand is usually given implicitly as a value function of some parametric optimization problem, strongly complicates the integration with respect to the probability measure and the optimization of the associated expectation functional. The main results of this work have been derived with the intention to ease these difficulties, and are seen in chapters 2, 3.

After introducing stochastic two-stage programs in chapter 1 we investigate duality relations for convex optimization problems with supply/demand and prices treated as parameters. Based on these results we state a stability criterion proving subdifferentiability of the value function under proper assumptions (chapter 2). In chapter 3 we derive bilinear approximates for the integrand of the underlying two-stage program with respect to specially shaped polytopes, socalled 'cross-simplices' (x-simplices). Employing the stability criterion we further prove the existence of bilinear functions, which minorize (majorize) the integrand optimally with respect to the expectation value. These minorants (majorants) support the integrand at generalized barycenters of the simplicial faces of the involved x-simplices, and amount to an approach which is denoted barycentric approximation scheme. After presenting an illustrative survey of existing approaches (chapter 4) we discuss issues on the implementation of the barycentric approximation scheme (chapter 5) and the gained experiences in stochastic two-stage programming (chapter 6).

This work was accepted as Habilitationsschrift by the Faculty of Economics of the University of Zurich in May 1992.

I should like to thank Prof. P. Kall for having given me the opportunity to work in his Institute, for introducing me to the area of stochastic programming and for providing me with facilities allowing efficient and fast implementation. I am especially grateful for his understanding and patience and for the generous freedom I received to pursue my own ideas in research, in teaching as well as in applications.

I owe thanks to Prof. S.M. Robinson, an anonymous referee and Prof. P. Kall for numerous valuable comments and remarks on this work.

I also benefited enormously from numerous conversations with Dr. R. Reinhardt and Dr. J. Mayer which helped me to clarify concepts. Further, I should like to thank Dr. J. Mayer and Dipl. math. E. Keller for their unstinted collaboration, in particular when I was working on the implementation of the barycentric approximation scheme. I have also to acknowledge the linguistic support I received by Miss. G. Utzinger.

Finally, I should like to thank Dr. Werner A. Müller from Springer-Verlag for the excellent cooperation.

Zurich, August 1992 Karl Frauendorfer

CONTENTS

0 PRELIMINARIES

In this part we state *definitions* (and *notations*) and *conventions* having been used throughout this work.

i) DEFINITIONS and NOTATIONS

We start with recalling known definitions and characterizations within linear topological vector spaces, (stochastic) (convex) analysis, measure and probability theory. For that which is outlined below, we mainly used Feller 1966 [44], Billingsley 1968 [9], Rockafellar 1970 [118] and 1976 [121], Halmos 1974 [55], Kelley and Namioka 1976 [80], Ekeland and Temam 1976 [39], Castaing and Valadier 1976 [20], Brøndsted 1983 [18], Prékopa 1980 [107], Rudin 1980 [131], Robinson 1987 [111], Wets 1989 [144] and Bauer 1990 [3] .

A **topology** for a set T is a family $\mathcal{O}$ of subsets of T, such that the empty set and T itself, the union of arbitrarily many sets of $\mathcal{O}$, the intersection of finitely many sets are in $\mathcal{O}$. The elements of $\mathcal{O}$ are called **open** sets. The set T with the topology $\mathcal{O}$ is denoted **topological space**. A subset X of T is **closed** iff the complement $T\setminus X$ is open. The **closure** of an arbitrary X of T, denoted **cl X**, is the intersection of all closed sets which contain X. A subset X of T whose closure contains T is **dense** in T. A topological space T is **separable** iff there exists a countable set X which is dense in T. The union of all open sets contained in X is an open set and called the **interior** of X, denoted **int X**. The **boundary** of a set X is **bd X**:= clX \ intX. A set U in T is a **neighborhood** of a point x in T iff x belongs to int U. The family $\mathcal{N}(x)$ of neighborhoods of x is called **neighborhood system** of x.

R denotes the real number system and $\mathbf{R}^n$ the **Euclidean space**, the vector space of real n-tuples. A function $f:\mathbf{R}^n \to \mathbf{R}^m$ is **linear** iff (if and only if) $f(\mu x+\lambda y)=\mu.f(x)+\lambda.f(y)$ for any $x,y \in \mathbf{R}^n$ and $\mu,\lambda \in \mathbf{R}$; in this case f is described with a matrix A of size $m\times n$; $A_{i,.}$ denotes the i-th row and $A_{.j}$ the j-th column. A function $f:\mathbf{R}^n \to \mathbf{R}$ is **positively homogeneous** iff for all $x \in \mathbf{R}^n$: $f(\mu x)=\mu.f(x)$ whenever μ is positive; f is **absolutely homogeneous** iff for all $x \in \mathbf{R}^n$: $f(\mu x)=|\mu|.f(x)$; f is **subadditive** iff $f(x+y) \leq f(x)+f(y)$. A nonnegative function on

$\mathbf{R}^n$ which is absolutely homogeneous, subadditive and takes the value 0 only at $0\in\mathbf{R}^n$ is called **norm**. A nonnegative function d(x,y) defined for each pair of points x,y in $\mathbf{R}^n$ is a **metric** iff d(x,x)=0, d(x,y)=d(y,x), d(x,y)≤d(x,z)+d(z,y), and d(x,y)=0 implies x=y. The **inner product** of two vectors x and x* in $\mathbf{R}^n$ is expressed by $<x,x^*>=x_1x_1^*+...+x_nx_n^*$. The inner product induces the **Euclidean norm** $\|x\|:=<x,x>^{1/2}$ and the **Euclidean metric** $d(x,x'):=\|x-x'\|=<x-x',x-x'>^{1/2}$ between two points x and x' in $\mathbf{R}^n$. The sum $\lambda_1x^1+...+\lambda_nx^n$ is called a **linear combination** of the vectors $x^1,...,x^n$. In the case that λ_i (i=1,...,n) are nonnegative one speaks of a **positive linear combination**; if $\lambda_i\geq0$ (i=1,...,n) add up to 1 we have a **convex linear combination** of the vectors $x^1,...,x^n$. A **basis** of $\mathbf{R}^n$ is a collection $\mathcal{B}$ of n linearly independent vectors.

The usual topology of the separable Euclidean space $\mathbf{R}^n$ is generated by the open balls $B_r(y)=\{x|\ \|x-y\|<r\}=\{x|d(x,y)<r\}$ of radii r (with arbitrary center y and r any rational number).

A subset X of $\mathbf{R}^n$ is said to be **bounded** iff there exists a positive r with $B_r(0)\supset X$. X is said to be **compact** iff X is bounded and closed. Let u,v be two points in $\mathbf{R}^n$: u,v are called **endpoints** of the **line segment** denoted by $[u,v]:=\{(1-\lambda)u+\lambda v \mid 0\leq\lambda\leq1\}$. X is **affine** iff $(1-\lambda)x+\lambda x'\in X$ whenever $x,x'\in X$ and $\lambda\in\mathbf{R}$; the smallest affine set containing X is called the **affine hull** of X denoted **aff X**. Affine sets that contain the origin are subspaces of $\mathbf{R}^n$. The **dimension of a non-empty affine set** is defined as the dimension of the subspace parallel to it. A set of m+1 points $\{a_0,...,a_m\}$ are said to be **affinely independent** iff aff$\{a_0,...,a_m\}$ is of dimension m; then aff$\{a_0,...,a_m\}$ is called **m-dimensional**. In this case any x in aff$\{a_0,...,a_m\}$ is uniquely representable through $x=\lambda_0a_0+...+\lambda_ma_m$ with $\lambda_0+...+\lambda_m=1$; herewith $\{\lambda_0,...,\lambda_m\}$, as parameters, is called a **barycentric coordinate system** for aff$\{a_0,...,a_m\}$. X is said to be **convex** iff $(1-\lambda)x+\lambda x'\in X$ whenever $x,x'\in X$ and $\lambda\in[0,1]$. The smallest convex set containing X is called the **convex hull** of X denoted **co X**. The **dimension of a convex set** is meant to be the dimension of its affine hull. The **relative interior** of a convex set X is **riX**$:=\{x|\exists r>0, B_r(x)\cap(\text{aff } X)\subset X\}$. A point $e\in X$ is called **extreme point** of a set X iff e is not in the relative interior of any line segments [x',x"] within X; the set of extreme points of X is denoted **ext-X**. For any nonzero $a\in\mathbf{R}^n$ the sets $\{x|<x,a><\alpha\}$ and $\{x|<x,a>\leq\alpha\}$ are called **open** and **closed half-spaces**. A subset X of $\mathbf{R}^n$ is a **polyhedral set** iff X can be expressed as an intersection of finitely many closed half-spaces of $\mathbf{R}^n$. X is a **polytope** iff it is the convex hull of finitely many points. Let X be a polytope: then a convex subset F of X is called a **face** iff for any two distinct points u,v in X such that any open line segment denoted]u,v[and $]u,v[\cap F$ are non-empty, we actually have $[u,v]\subset F$. The 1-dimensional faces of a polytope are called **edges**. The extreme points of a polytope are called **vertices**. Two distinct vertices of a polytope are called **adjacent** iff the line segment joining them is an edge. It is said that a vertex and an edge are **incident** iff the vertex is a vertex of the edge. A polytope X of dimension n is called **simple** iff each vertex of X is incident to precisely n edges. Iff $\{a_0,...,a_m\}$ are affinely independent, its convex hull is called a **simplex** (which in turn is simple). The point $\lambda_0 a_0+...+\lambda_m a_m$ with $\lambda_0=...=\lambda_m=1/(1+m)$ is called **midpoint** or **barycenter** of the simplex. Any point in a simplex is uniquely given through its non-negative barycentric coordinates; non-negative barycentric coordinates shall be called **barycentric weights**. A simplex in $\mathbf{R}^n$ with dimension n is called **regular** .

Let $\{C^v, v=0,1,2,...\}$ be a sequence of sets in $\mathbf{R}^n$. The **limit inferior** of $\{C^v, v=0,1,2,...\}$ is

$$\textbf{lim inf}_{v\to\infty} C^v := \{x \mid x=\lim_{v\to\infty} x^v,\ x^v\in C^v \text{ for } v=0,1,2,...\};$$

points of this set are called **limit points**. The **limit superior** of $\{C^v, v=0,1,2,...\}$ is

$$\textbf{lim sup}_{v\to\infty} C^v := \{x \mid x=\lim_{k\to\infty} x^{v_k}, x^{v_k}\in C^{v_k} \text{ for some subsequence } \{v_k, k=0,1,2,...\}\subset\{0,1,2,...\}\};$$

points of this set are called **accumulation points**. The limit $\textbf{lim}_{v\to\infty} C^v$ exists iff $\lim\inf_{v\to\infty} C^v = \lim\sup_{v\to\infty} C^v$.

The function f is said to be **real-valued** iff f takes values in **R**. Let f and g denote real-valued functions defined on $\mathbf{R}^n$; the socalled **Landau symbols** o(g(x)) and O(g(x)) are defined according to : **f(x)=o(g(x)) (x→0)** iff $\lim_{x\to 0} |f(x)|.|g(x)|^{-1}=0$ and **f(x)=O(g(x)) (x→0)** iff there exists a constant C so that $|f(x)|\le C.|g(x)|$ in a proper neighborhood of 0. The function f is said to be **bounded** on a subset X of $\mathbf{R}^n$ iff there exist two numbers α' and α'' for which $\alpha'\le f(x)\le\alpha''$ for all x on X; f is said to be **extended-real-valued** iff f takes values in $\mathbf{R}\cup\{-\infty,+\infty\}$. $\mathbf{f^+(x)}:=\max\{0,f(x)\}$ and $\mathbf{f^-(x)}:=\min\{0,f(x)\}$. I_X denotes the **indicator function** of X ($I_X(x):=0$ for $x\in X$, $I_X(x):=+\infty$ for $x\notin X$).
Let f denote an extended-real-valued function on $\mathbf{R}^n$: the set $\{(x,\alpha) \mid x\in\mathbf{R}^n, \alpha\in\mathbf{R}, \alpha\ge f(x)\}$ is called **epigraph** of f and denoted **epi f**; the **hypograph** of f is denoted with **hyp f**$:=\{(x,\alpha) \mid x\in\mathbf{R}^n, \alpha\in\mathbf{R}, \alpha\le f(x)\}$. The **effective domain** of f is **dom**

f:=$\{x|f(x)<\infty\}$, i.e. the projection of the epigraph on $\mathbf{R}^n$; f is said to be **proper** iff f takes nowhere the value $-\infty$ and is not identically $+\infty$; otherwise the function f is called **improper**. $(f\oplus g)(x)=\inf\{\alpha \mid (x,\alpha) \in (\text{epi } f + \text{epi } g)\}$ denotes the **infimal convolution** (defined for proper functions) or denotes the **addition of epi-graphs** (defined for arbitrary functions) of two functions f and g. A function whose epigraph covers epi f is called a **minorant** of f; a function whose epigraph is a subset of epi f is called a **majorant** of f. The **α-level sets** of a function f are given by $\text{lev}_\alpha f:=\{x|f(x)\leq\alpha\}$; f is said to be **inf-compact** iff all the level-sets are bounded. The set of **minimizers** of f is **argmin f**:= $\{x'|f(x')=\inf_x f(x)\}$; the set of ε-minimizers is **ε-argmin f** := $\{x'|f(x')\leq\inf_x f(x)+\varepsilon\}$. A nonempty subset M of X is called a **complete local minimizing set** for f with respect to an open set $G\supset M$, iff the set of minimizers of f on clG is M. The function f is said to be **lower semicontinuous (lsc)** iff epi f is closed; the greatest lower semicontinuous function majorized by f is called the **lower semicontinuous hull** of f; f is **upper semicontinuous (usc)** iff epi [-f] is closed; if f is both lower and upper semicontinuous then f is said to be **continuous**. A real-valued function f is called **Lipschitzian** on a set X iff there exists a nonnegative α so that $|f(x)-f(x')|\leq\alpha|x-x'|$ for any x,x' in X; in the case that f is Lipschitzian in a neighborhood of x' then f is said to be **locally Lipschitzian** at x'; α is called **Lipschitz constant**; f is **polyhedral** (or **piecewise linear**) iff epi f is a polyhedral set.

The function f is said to be **convex** iff epi f is a convex subset of $\mathbf{R}^{n+1}$. A real-valued function f on X is **strictly convex** iff $f(\lambda.x+(1-\lambda).y))<\lambda.f(x)+(1-\lambda)f(y)$ for any $x\neq y$ in X and $0<\lambda<1$. A function f is called **sublinear** iff f is convex and positively homogeneous; f is **concave** iff -f is convex (so, all characterizations outlined in this paragraph for convex functions can easily be stated for concave ones via -f). A function is said to be **linear affine** on X iff f is finite, convex and concave on X. The function f is **proper convex** if epi f is nonempty, convex and does not contain a vertical line; else f is said to be **improper convex**. In the case that a function f may be written as $f(x)=\langle x,Qx\rangle+\langle c,x\rangle$ with Q a nxn matrix and $c\in\mathbf{R}^n$ then f is called **linear-quadratic;** the matrix Q is said to be **positive definite** iff $\langle x,Qx\rangle$ is positive for any nonzero x; iff $\langle x,Qx\rangle$ is nonnegative for any x then Q is said to be **positive semi-definite**. A linear-quadratic function is convex iff Q is positive semi-definite.
The **closure** of a proper convex function, denoted **cl_x f** (or **cl f** if there is no danger of confusion) is defined to be the lower semicontinuous hull of f. With A being a linear transformation from $\mathbf{R}^n$ to $\mathbf{R}^m$ the function $(Af)(y)=\inf\{f(x)|Ax=y\}$ is called the **image of f under A**; (Af)(y) is convex if f(x) is convex. A proper convex function f is called **subdifferentiable** at x', iff f has a linear minorant

coinciding with f at x'; the slope of that minorant is called **subgradient**; the set of subgradients of f at x' is called the **subdifferential at x'** and denoted $\partial f(x')$. If the subdifferential at x' is a singleton (i.e. consists of exactly 1 element) then f is said to be **differentiable at x'**; if a convex function f is differentiable and finite throughout $\mathbf{R}^n$ then f is called **smooth**.

Let f be a convex function: the **conjugate** function of f is denoted f* and defined $f^*(x^*)=\sup_x \{<x,x^*>-f(x)\}$; the conjugate f* is then a closed convex function, proper iff f is proper; moreover, $(\mathrm{cl}\, f)(y)=\sup_{x^*} \inf_x\{<y-x,x^*>+f(x)\}$ for any y, and $(\mathrm{cl}\, f)^*=f^*$, $f^{**}=\mathrm{cl}\, f$.
Let f be a concave function: the **conjugate** function of f is denoted f* and defined $f^*(x^*)=\inf_x \{<x,x^*>-f(x)\}$; the conjugate f* is then a closed concave function, proper iff f is proper; moreover, $(\mathrm{cl}\, f)(y)=\inf_{x^*} \sup_x\{<y-x,x^*>+f(x)\}$ for any y, and $(\mathrm{cl}\, f)^*=f^*$, $f^{**}=\mathrm{cl}\, f$.
The function f(x,y) is called a **concave-convex saddle function** on $\mathbf{R}^{n+m}$ iff f(.,y) is concave for all $y\in \mathbf{R}^m$ and f(x,.) is convex for all $x\in \mathbf{R}^n$. The subdifferential of a saddle function f(.,.) at (x',y') denoted $\partial f(x',y')$ is defined: $\partial f(x',y')=\partial_x f(x',y')\times \partial_y f(x',y')$. A point (x',y') is called **saddle point** of a saddle function f(.,.) iff $\partial f(x',y')$ includes the origin. Such a point is also called **stationary**.

The collection $\{f^\nu, \nu=1,2,....\}$ of real-valued functions **converge uniformly** to f on a subset X of $\mathbf{R}^n$ iff for any positive ε there exists a $N(\varepsilon)$ so that $|f^\nu(x^\nu)-f(x)|<\varepsilon$ $\forall x\in X$, $\forall \nu\geq N(\varepsilon)$. Uniform convergence on X implies **pointwise convergence** on X (i.e. $\lim_{\nu\to\infty} f^\nu(x)=f(x)$ for any x in X). A collection $\{f^\nu, \nu=1,2,....\}$ of real-valued functions on $\mathbf{R}^n$ converge to f relative to **the uniform convergence on compact sets** iff the collection converges uniformly on every compact set in $\mathbf{R}^n$. Let $\{f(p,x)|p\in \Pi\}$ denote a set of real-valued functions in x on $\mathbf{R}^n$ with Π denoting an arbitrary set: $\{f(p,x)|p\in \Pi, x\in X\subset \mathbf{R}^n\}$ is said to be **uniformly equicontinuous** on X iff for an arbitrary positive ε there exists a δ such that $|f(p,x)-f(p,y))|<\varepsilon$ for any p in Π and $\|x-y\|<\delta$. Let Π be some topological space: the **epi-limit superior** of f at x' as $p\to p'$, denoted $(E^*f)(x')$, and the **epi-limit inferior** of f at x' as $p\to p'$, denoted $(E_*f)(x')$, are defined

$$(E^*f)(x') := \sup_{V\in \mathcal{N}(x')} \limsup_{p\to p'} \inf_{x\in V} f(p,x)$$
$$(E_*f)(x') := \sup_{V\in \mathcal{N}(x')} \liminf_{p\to p'} \inf_{x\in V} f(p,x) ;$$

f is **epi-upper semicontinuous at x' as $p\to p'$** iff $f(p',x')\geq (E^*f)(x')$ and **epi-lower semicontinuous at x' as $p\to p'$** iff $f(p',x')\leq(E_*f)(x')$; f is **epi-continuous at x' as $p\to p'$** iff f is both there.

Denoting with Ω,X arbitrary sets then a **multifunction** $\Gamma:\Omega\to X$ is a subset of $\Omega\times X$. The set of all $x\in X$ such that $(\omega,x)\in\Gamma$ is denoted $\Gamma(\omega)$. The **domain** and **graph** of Γ are **dom** $\Gamma:=\{\omega\in\Omega|\ \Gamma(\omega)\neq\emptyset\}$ and **gph** $\Gamma:=\{(\omega,x)|x\in\Gamma(\omega)\}$. In the case that X is equipped with a topology then Γ is said to be **closed-valued** iff $\Gamma(\omega)$ is a closed set for all $\omega\in\Omega$. Γ is **convex-valued** if $\Gamma(\omega)$ is a convex set for all $\omega\in\Omega$. If both Ω and X are equipped with a topology then Γ is called **Berge upper semicontinuous** iff for any open set W with $W\supset\Gamma(\omega')$ there exists a neighborhood $U\in\mathcal{N}(\omega')$ so that $\Gamma(\omega)\subset W$ for all $\omega\in U$.

A **σ-field** (**σ-algebra**) for a set Ω is a family $\mathcal{M}$ of subsets of Ω, such that Ω itself, the complement of any set of $\mathcal{M}$, and the union of countably many sets of $\mathcal{M}$ are in $\mathcal{M}$. The elements of $\mathcal{M}$ are called **measurable** sets. The set Ω with the σ-field $\mathcal{M}$ is called a **measurable space** being denoted $(\Omega,\mathcal{M})$. The **product-σ-field** of two measurable spaces, say $(\Omega^1,\mathcal{M}^1),(\Omega^2,\mathcal{M}^2)$, on $\Omega^1 x\Omega^2$ is the smallest σ-field, denoted $\mathcal{M}^1 x\mathcal{M}^2$, containing all $M^1\times M^2$ with M^1,M^2 being measurable. In the case that Ω is equipped with a topology then the **Borel field** for Ω is the smallest σ-field containing all the compact subsets of Ω. A **measure** μ on $(\Omega,\mathcal{M})$ is a nonnegative, real-valued set function which is countably additive (i.e.: $\mu(\cup_i M_i)=\Sigma_i\ \mu(M_i)$ for every sequence $\{M_i,\ i=1,2,...\}$ of disjoint, measurable sets), and assigns 0 to the empty set. $(\Omega,\mathcal{M},\mu)$ is called **measure space**. The **Lebesgue measure** in $\mathbf{R}^n$ is uniquely determined on its Borel field $\mathcal{B}^n$ through $\lambda(\{x|a<x\leq b\}=\Pi_i(b_i-a_i)$ for $a_i<b_i$ $(i=1,...,n)$.

Let Ω be a linear vector space equipped with a topology and a σ-field $\mathcal{M}$. A **probability measure (probability distribution)** P on $(\Omega,\mathcal{M})$ is a measure that assigns 1 to the entire set Ω. $(\Omega,\mathcal{M},P)$ is called **probability space**. A probability measure is **logarithmically concave** iff for every A,B and $\{\lambda A+(1-\lambda)B\mid 0<\lambda<1\}$ of measurable convex subsets of Ω it holds that:

$$P(\lambda A+(1-\lambda)B)\geq[P(A)]^{\lambda}.[P(B)]^{(1-\lambda)}$$

for every $0<\lambda<1$ (where '+' means Minkowski addition of sets, i.e. $A+B=\{a+b|a\in A,b\in B\}$ and the constant multiple of a set is defined as $\lambda A=\{\lambda a|a\in A\}$ for every fixed real number λ). A proposition holds **P-almost surely (P-a.s.)**, if it holds on a subset A of Ω with P(A)=1. Let B be a measurable set in Ω with positive probability P(B). Then, the corresponding **conditional probability measure** P^B on $(\Omega,\mathcal{M})$ is given with $P^B(M)=P(M\cap B).[P(B)]^{-1}$ for all measurable M. The **support** of $(\Omega,\mathcal{M},P)$, denoted **supp P**, is the smallest closed set in Ω with probability 1. In the case that the support is representable through a countable union of points, the probability measure is said to be **discrete**. Points in Ω with positive probability are called **atoms**. Let $(\Omega^1,\mathcal{M}^1,P^1)$ and $(\Omega^2,\mathcal{M}^2,P^2)$ denote two probability spaces. Then P generated by $P(M^1 \times M^2)=P^1(M^1).P^2(M^2)$ is called the **product measure** on $(\Omega^1\times\Omega^2,\mathcal{M}^1\times\mathcal{M}^2)$. To the contrary let $(\Omega^1\times\Omega^2,\mathcal{M}^1\times\mathcal{M}^2)$ be a product space equipped with a probability measure P; then P induces probability measures P^1 and P^2 on $(\Omega^1,\mathcal{M}^1)$ and $(\Omega^2,\mathcal{M}^2)$, through: $P^1(M^1)=P(M^1\times\Omega^2)$ and $P^2(M^2)=P(\Omega^1\times M^2)$ with M^1,M^2 measurable; P^1,P^2 are called **marginal distributions**. With $(\Omega,\mathcal{M},P)$, a function $y:\Omega\to\mathbf{R}^n$ is called **measurable** iff $y^{-1}(B):=\{\omega|y(\omega)\in B)\}$ is measurable for any Borel set B of $\mathbf{R}^n$; in this case y is a **random vector**; a random vector y induces a probability measure P^y on $\mathbf{R}^n$ through $P^y(B)=P\{y^{-1}(B)\}$ for any Borel set B of $\mathbf{R}^n$. The random vectors $y^i:\Omega\to\mathbf{R}^{n_i}$ (i=1,...,I) are **stochastically independent** iff $P\{\omega|y^i(\omega)\in B^i,\ i=1,...,I\}=\Pi_i P\{\omega|y^i(\omega)\in B^i\}$ for arbitrary Borel sets B^i (i=1,...,n) of $\mathbf{R}^{n_i}$. This is the case iff P^y induced by $y=(y^1,...,y^I)$: $\Omega\to\times_i\mathbf{R}^{n_i}$ on $(\times_i\mathbf{R}^{n_i},\times_i\mathcal{B}^{n_i})$ is equal to $\times_i P^{y,i}$ with $P^{y,i}$ being induced by y^i: $\Omega\to\mathbf{R}^{n_i}$ on $(\mathbf{R}^{n_i},\mathcal{B}^{n_i})$ (i=1,...,I). P^y on $(\mathbf{R}^n,\mathcal{B}^n)$ is **absolutely continuous** with respect to the Lebesgue measure iff $\lambda(B)=0$ implies $P^y(B)=0$. P^y is completely defined through its **distribution function** $F(a):=P^y\{y|y\le a\}$; distribution functions corresponding to marginal probability distributions are called **marginal distribution functions**.

Let f denote a measurable function on the probability space $(\mathbf{R}^n,\mathcal{B}^n,P)$; $\int f dP$ denotes the **integral** of f with respect to P; f is **integrable** with respect to P iff both $\int f^+ dP$ and $\int f^- dP$ are finite; the **integral exists** (f is **quasi-integrable** with respect to P) iff either $\int f^+ dP$ or $\int f^- dP$ is finite. A sequence $\{P_j\ ,\ j=1,2,...\ \}$ of probability measures on $(\mathbf{R}^n,\mathcal{B}^n)$ **converges weakly** to P iff for all bounded continuous functions f on $\mathbf{R}^n$ it holds that $\lim_{j\to\infty}\int f dP_j = \int f dP$. Let F denote the distribution function on $\mathbf{R}^n$: then $\int f dF$ is called **Lebesgue-Stieltjes integral**. Let ω denote a random vector with values in $\mathbf{R}^n$ and induced probability measure P on $(\mathbf{R}^n,\mathcal{B}^n)$: the integral $[\ \int \omega_i\, dP(\omega)\ ]$ is called the **expectation** of ω_i with respect to P; more generally, we shall call $[\ \int \Pi_{i\in\Lambda}\,\omega_i\, dP(\omega)\ ]$ a **cross moment** of ω for any subset $\Lambda\subset\{1,...,n\}$. Further, let $\{\rho(.,x):\mathbf{R}^n\to\mathbf{R}|\ x\in X\}$ denote a collection of measurable and integrable functions w.r.t. P, then $(E\rho)(x):=\int\rho(\omega,x)dP(\omega)$ is called **expectation functional** on X.

Let Ω be equipped with a Borel field $\mathcal{B}$ and $\mathbf{R}^n$ with its usual topology. The multifunction Γ: $\Omega\to\mathbf{R}^n$ is called **measurable** with respect to (w.r.t.) $(\Omega,\mathcal{B})$ iff for any open set B in $\mathbf{R}^n$ the set $\Gamma^{-1}(B):=\{\omega\in\Omega|\Gamma(\omega)\cap B\neq\emptyset\}$ is a measurable subset of Ω. Let P denote a probability measure on $(\Omega,\mathcal{B})$. A function $y:\mathrm{dom}\Gamma\to\mathbf{R}^n$ is called a **selection** (associated with Γ) iff $y(\omega)\in\Gamma(\omega)$ for any $\omega\in\mathrm{dom}\Gamma$. The expectation of a measurable multifunction with respect to P is the set :

$$\int\Gamma(\omega)\,\mathbf{dP}(\omega) := \mathrm{cl}\ \{\ \int y(\omega)\ dP(\omega) \mid y(.) \text{ is a measurable and integrable selection}\}\ .$$

An extended-real-valued function $f{:}\Omega\times\mathbf{R}^n \to \mathbf{R}\cup\{-\infty,+\infty\}$ is called an **integrand** on $\Omega\times\mathbf{R}^n$; f is a **lower semicontinuous integrand** iff $f(\omega,x)$ is lower semicontinuous in x for every $\omega\in\Omega$; f is a **normal integrand** iff f is lower semicontinuous and epi $f(\omega,.)$ is a measurable multifunction w.r.t. $(\Omega,\mathcal{B})$; f is a **proper integrand** iff $f(\omega,.)$ is proper for every $\omega\in\Omega$; f is a **convex integrand** iff $f(\omega,.)$ is convex for every $\omega\in\Omega$; f is called a **Carathéodory integrand** iff $f(\omega,x)$ is finite, measurable in ω, and continuous in x.

ii) CONVENTIONS

Based on the above outlined definitions we below introduce new notions and state conventions.

Viewing a **simplex** in $\mathbf{R}^K$ as a special polytope, then a simplex is **closed**. In this work we shall allow for closed, open and half-open simplices: a simplex is **open** iff it is the intersection of those open (finitely many) halfspaces whose closure characterize the closure of the simplex (again through intersection); a simplex is **half-open** if it is the intersection of those open and closed halfspaces, whose closure characterize the closure of the simplex (again through intersection). Open and half-open simplices are called **regular** iff their closure is regular. In this sense open, closed and half-open simplices are **measurable** (w.r.t. the Borel field) as these are characterized as intersections of finitely many open and closed (and hence measurable) halfspaces.

In addition to the above we shall introduce the notion $\times$-**simplex** (say cross-simplex) for 'specially shaped' polytopes (in the case of closedness) : $\Xi\subset\mathbf{R}^K$ will be called a $\times$-**simplex** in $\mathbf{R}^K$ iff Ξ is representable as Cartesian product of finitely many simplices; i.e.

$$\Xi = \mathop{\mathrm{X}}_{i=1}^{I} \Xi^i$$

with Ξ^i (i=1,...,I) being simplices of $\mathbf{R}^{K_i}$. A ×-simplex Ξ is said to be **closed** or **open** iff all simplices Ξ^i are closed or open. A ×-simplex Ξ is said to be **half-open** iff either some simplices Ξ^i are half-open, or Ξ is characterized as a Cartesian product of open and closed simplices. Moreover, we shall say that Ξ **is regular** iff Ξ^i are regular simplices in $\mathbf{R}^{K_i}$ (i=1,...,I) and $K_1+...+K_I = K$.

Finally we shall say that a ×-simplex Ξ is **compatible with stochastic independence of** the probability space

$$\mathop{\mathrm{X}}_{i=1}^{I} (\mathbf{R}^{K_i}, \mathcal{B}^{K_i}, P^i)$$

iff Ξ may be represented as

$$\Xi = \mathop{\mathrm{X}}_{i=1}^{I} \Xi^i$$

with Ξ^i being ×-simplices and supp $P^i \subset \Xi^i \subset \mathbf{R}^{K_i}$.

Remark: ×-simplices may be seen as generalizations of (multidimensional) rectangles that obviously are Cartesian products of intervals (one-dimensional simplices). Moreover ×-simplices have the nice property that one may easily evaluate the number of vertices, the number of edges (one-dimensional faces) and the edges itself, due to the property that any two vertices of a simplex are adjacent. Further, it is easy to prove that ×-simplices belong to the class of *simple polytopes*.

I Stochastic Two-Stage Problems

Optimization problems, in particular mathematical programs, are characterized by a *set of* (in particular finite dimensional) *alternatives (decisions)* and an *objective*. Decision makers have to select out of this set that alternative (decision), best satisfying the underlying objective. If all data are known prior to the decision process we speak of *deterministic optimization (programming)*; if part of the data remain unknown and available only in a certain probabilistic sense we speak of *stochastic optimization (programming)*. We retain on modeling uncertain data by means of assigning probability distributions to these data. Hence, we take as a premise that a probability distribution is given.

Being familiar with convex analysis we are well aware of the importance of convexity in the field of optimization. This is due to the Fenchel-Rockafellar duality theory developed in Rockafellar 1970 [118], 1974 [119], that helps to understand and solve optimization problems. This theory tells us that a dual pair of convex optimization problems correspond one-to-one with minimax problems. More precisely, the dual arises from embedding the given problem in a family of problems by means of convex parameterization. The so obtained parametric convex optimization problem entails an optimal value funtion. This function and the dual problem reveals a variety of relations. Based on these relations highly sophisticated methods have been developed that are fast, stable, exploit convexity and solve convex, smooth mathematical programs accurately (see e.g. Bazaraa and Shetty 1979 [4], Fletcher 1981 [45], Gill, Murray and Wright 1981 [54], Bertsekas 1982 [8]). For deterministic problems the challenge nowadays lies rather in nonsmooth optimization (see Clarke 1989 [22] and references therein) as well as in nonconvex optimization (see Horst and Tuy 1990 [61] and references therein). The state-of-the-art within optimization is completely different as soon as uncertain data are incorporated : stochastic linear optimization problems are still challenging. Within stochastic optimization, one faces besides minimization or maximization the onerous task of integrating multidimensional functions which are often available only implicitly; additionally, these functions may depend on decisions, causing enormous difficulty in even judging the goodness of a decision, not to speak of how to deduce the best possible one. It is this fact that has caused an enormous gap between stochastic models introduced up to now in literature and existing (numerically tractable) methods for solving them.

Stochastic programming has its origins in the mid 50s with the works of Beale 1955 [5], Dantzig 1955 [23], Tintner 1955 [138], Charnes and Cooper 1959 [21]. Based on these works, essentially two types of modeling turned out and attained considerable attention since then: *Stochastic Two-Stage Programming* (to which this work mainly contributes) and *Programming under Probabilistic Constraints*.

Within *Programming under Probabilistic Constraints* the decision is chosen prior to the outcome of the random data and with respect to a set of decisions that satisfy constraints (or targets, goals) with a prescribed reliability level. Considerable advance with respect to modeling as well as solving probabilistic constrained programs has been achieved with the theory of *logarithmically concave probability measures* developed by Prékopa 1973 [105], 1974 [106]. For a summarizing paper of that theory we refer to Prékopa 1980 [107]. Solution techniques (based on these investigations) are described in Prékopa 1988 [108] and in references cited therein.

For an overview and history of *Stochastic Programming* in general, we refer to Dempster 1980 [30] and 1986 [31], Wets 1983 [142], Ermoliev and Wets 1988 [43]. Challenging extensions of the above-mentioned models may be found in Ermoliev and Wets 1988 [42], Wets 1989 [144], Birge and Wets 1991 [17].

The importance of stochastic programming in practical life is seen immediately in thinking of e.g. inventory problems, assets management, production planning, where deciding prior to available data on demand and prices is a necessity. Based on this fact, additional costs might arise which could be reduced considerably, if sophisticated analysis of uncertainty with respect to the underlying model is performed. For descriptions of practical problems within stochastic programming we refer to Sengupta 1972 [135], Dempster 1980 [29], Ermoliev & Wets 1988 [42], and Birge and Wets 1991 [17].

This work contributes to investigations within *Stochastic Two-Stage-Programming* (as the title indicates). In particular, throughout this chapter (I) we concentrate on modeling and on structural properties of this model type. The main result of this work is seen in chapter III, where we develop the socalled barycentric approximation scheme. This allows to approximate multidimensional integrals by expressions of at most two-dimensional integrals. These integrals refer to bilinear terms defined on $\times$-simplices being compatible with stochastic independence of random data. In particular for a rather general class of stochastic two-stage programs we analyse the goodness of barycentric approximations and refine these appropriately, based on duality results achieved in chapter II. In chapter IV we give an illustrative survey of solution techniques designed to solve stochastic two-stage programs. Implementation issues of the barycentric approximation scheme are treated in chapter V. In chapter VI we apply our approach to stochastic linear two-stage problems and report on computational results and experiences.

Throughout this chapter we shall focus on modeling stochastic two-stage problems (sections 1 and 2). Based on well-established concepts, we therein state conditions that guarantee stochastic (convex) two-stage problems to be well-defined. Section 3 outlines recent stability results, being of importance in the case problem data might be perturbed. Epi-convergence will be discussed in section 4, intending to illustrate the criteria with respect to which approximation schemes ought to to be designed. In sections 5 and 6 we raise the saddle property within stochastic two-stage programs, taking into consideration stochastic independences of random data. In section 7 we outline special convex cases of stochastic two-stage problems, intending to discuss well-known features that ought to be exploited by solution methods.

Let ω be a random vector with values in Ω, a closed subset of $\mathbf{R}^M$, being equipped with the Borel field. The induced probability measure is P. Thus, let $(\Omega,\mathcal{B}^M,P)$ denote the underlying probability space, where $\mathcal{B}^M$ is the set of Borel sets restricted to Ω.

We start with convex stochastic programs of the form

$$\inf_{x\in X} \quad f(x) + \int_{\Omega} \rho(\omega,x)dP(\omega) \quad . \tag{1.1}$$

X represents the set of feasible decisions and is supposed to be a convex, closed subset of $\mathbf{R}^n$; f: $X \to \mathbf{R}\cup\{\infty\}$ is proper convex, extended-real-valued. We assume that for any x in X :

(A1) $\quad \omega \to \rho(\omega.,x)$ is measurable w.r.t. $(\Omega,\mathcal{B}^M)$,

(A2) $\quad \omega \to \rho^-(\omega,x)$ is integrable w.r.t. P .

(A1) and (A2) ensure that the integral in (1.1) is well defined.

Remark 1.1: Throughout this work 'measurable', 'measurability' refers to the σ-algebra of Borel sets (Borel field) restricted to some closed subset of a finite-dimensional Euclidean space; of course, that subset may be the entire space. However, allowing such restrictions on measurability will increase the applicability of advantageous structural properties, which are meaningful for being exploited.

Denoting

$$(E\rho)(x) := \int_{\Omega} \rho(\omega,x)dP(\omega) \tag{1.2}$$

we get the following statement :

Proposition 1.1 (Wets 1989 [144, Propositions 2.1. and 2.2.]): Suppose the function ρ: $\Omega x X \to \mathbf{R}\cup\{\infty\}$ is convex in x, P-almost surely (P-a.s.). Then the expectation functional $(E\rho)(x)$ is convex in x on X; if additionally $\rho(\omega,x)$ is lower semicontinuous in x (P-a.s.), then $(E\rho)(x)$ is lower semicontinuous, too.

This yields a convex objective function $f(x)+(E\rho)(x)$ and hence a convex stochastic program (1.1). In special cases $(E\rho)(x)$ is easily computable: for example, if P is discrete with finite support (not too large) where integration becomes a finite summation, or if $\rho(.,x)$ is separable in the M components of ω entailing that $(E\rho)(x)$ is representable as a sum of M one-dimensional integrals. In these cases one might attack directly the socalled deterministic equivalent problem of (1.1)

$$\inf_{x\in X} \quad f(x) + (E\rho)(x) \tag{1.3}$$

with employing convex optimization methods. Usually one has to aim at a more sophisticated interplay between minimization and multidimensional integration.

An interpretation of (1.1) may simply be given: for finding a decision x within X prior to any information on the outcome ω, one seeks to find the best decision with respect to a desired objective, say minimizing costs. In our case f(x) represents the deterministic costs (only dependent on decision x) and $\rho(\omega,x)$ the (random) costs dependent on decision x and outcome ω. Integrating the latter with respect to P yields the average costs (s.t. x) to be expected. The total costs (i.e. deterministic and expected) are to minimize with respect to <u>no information</u> (except the information of probability distribution) about the outcomes.

The function $\rho(\omega,x)$ in (1.1) might be prescribed either explicitly (e.g. in form of an utility function) or implicitly in form of an optimal value function of an 'inner' convex optimization problem in the stage where outcomes are known. In the latter case we may write

$$\rho(\omega,x) = \inf_{y\in Y(\omega,x)} \phi(\omega,y,x) \tag{1.4}$$

where for every x in X, $Y(.,x)\colon \Omega\rightarrow\mathbf{R}^{n'}$ denotes a closed-convex-valued multifunction and $\phi(.,.,x)\colon \Omega\times\mathbf{R}^{n'} \rightarrow \mathbf{R}$ an extended-real-valued function.

With (1.4) we pass over to the well-known class of *convex stochastic two-stage problems* (also called *convex stochastic programs with recourse*) being of the form

$$\inf_{x\in X} \quad f(x) + \int_{\Omega} \inf_{y\in Y(\omega,x)} \phi(\omega,y,x)dP(\omega) \ . \tag{1.5}$$

Remark 1.2: Below we refrain from specifying assumption (A2) in more detail; integrability of $\rho^-(.,x)$ can be ensured through exploiting several characteristica of P, Ω, $\phi(\omega,y,x)$, $Y(\omega,x)$. Some of these will become apparent later (chapter III) when we specify the frame subject to which the barycentric approximation scheme is applied. In this section and the following one, we will focus on measurability of $\rho(\omega,x)$.

To ensure measurability (A1) we apply the concept of *proper normal convex integrands* (due to Rockafellar 1968 [116], 1969 [117] and 1976 [121]). We assume for every x in X:

(A1)' - $y \to \phi(\omega,y,x)$ is proper convex, lower semicontinuous on $\mathbf{R}^{n'}$, for every $\omega \in \Omega$;
- $\omega \to \phi(\omega,y,x)$ is measurable w.r.t. $(\Omega,\mathcal{B}^M)$ for every $y \in \mathbf{R}^{n'}$;
- int $\{y \mid \phi(\omega,y,x) < +\infty\} \neq \emptyset$ for every $\omega \in \Omega$;
- $\omega \to Y(\omega,x)$ is a closed-convex-valued multifunction, measurable w.r.t. $(\Omega,\mathcal{B}^M)$.

The first three conditions of (A1)' imply for every x in X, that $(\omega,y) \to \phi(\omega,y,x)$ is a proper normal convex integrand (Rockafellar 1969 [117, p.6]). This and the fourth condition in (A1)' ensure that the optimal value function $\omega \to \rho(\omega,x)$ is measurable ([117, Corollary 4.3]).
Finally, applying Rockafellar 1970 [118, p. 38] we get convexity of $\rho(\omega,x)$ in x on X (P-almost surely) with supposing $(y,x) \to \phi(\omega,y,x)$ to be convex in (y,x) on $\mathbf{R}^{n'} \times X$ (P-almost surely).

We mention that any parametric convex program of the form

$$\rho(\omega,x) := \inf q(\omega,y) \tag{1.6.1}$$
$$\text{s.t. } g_i^1(y) \leq h_i^1(\omega,x) \qquad i=1,\ldots,r_1 \tag{1.6.2}$$
$$g_j^2(y) = h_j^2(\omega,x) \qquad j=1,\ldots,r_2 \tag{1.6.3}$$

is of type (1.4) under the following assumptions supposed to hold for every $x \in X$:

(A1)'' - $y \to q(\omega,y)$ is proper convex, lower semicontinuous on $\mathbf{R}^{n'}$, for every $\omega \in \Omega$;
- $\omega \to q(\omega,y)$ is measurable w.r.t. $(\Omega,\mathcal{B}^M)$ for all $y \in \mathbf{R}^{n'}$;
- int $\{y \mid q(\omega,y) < +\infty\} \neq \emptyset$ for every $\omega \in \Omega$;
- $y \to g_i^1(y)$ is convex, real valued on $\mathbf{R}^{n'}$ $(i=1,\ldots,r_1)$;
- $y \to g_j^2(y)$ is linear on $\mathbf{R}^{n'}$ $(j=1,\ldots,r_2)$;
- $\omega \to h_i^1(\omega,x)$ is measurable w.r.t. $(\Omega,\mathcal{B}^M)$ $(i=1,\ldots,r_1)$;
- $\omega \to h_j^2(\omega,x)$ is measurable w.r.t. $(\Omega,\mathcal{B}^M)$ $(j=1,\ldots,r_2)$.

Analogously to the above, the first three conditions in (A1)" entail $q(\omega,y)$ to be a proper normal convex integrand. The assumptions on $g_i^1(.), g_j^2(.), h_i^1(.,x)$ and $h_j^2(.,x)$ yield a measurable, closed-convex-valued multifunction $\omega \rightarrow Y(\omega,x)$, where $Y(\omega,x)$ is the feasibility set of (1.6.2),(1.6.3) for given (ω,x) (Rockafellar 1969 [117, Corollary 1.3] and Rockafellar 1976 [121, Corollary 1Q]). This implies that the optimal value function $\omega \rightarrow \rho(\omega,x)$ of (1.6) is measurable ([117, Corollary 4.3]).

Convexity (P-a.s.) of $x \rightarrow \rho(\omega,x)$ is implied if $h_i^1(\omega,.)$ $(i=1,\dots,r_1)$, $h_j^2(\omega,.)$ $(j=1,\dots,r_2)$ are linear in x (P-a.s.) .

2 *Nonconvex Case*

Allowing nonconvexity of X, f(x) and $\rho(\omega,x)$ in x we obtain a stochastic program

$$\inf_{x \in X} \quad f(x) + \int_{\Omega} \rho(\omega,x) dP(\omega) \tag{2.1}$$

with now X being a closed subset of $\mathbf{R}^n$; the assumptions concerning probability space $(\Omega, \mathcal{B}^M, P)$ remain the same as for (1.1); again $\rho(.,x)$ is supposed to satisfy (A1), (A2) for any x in X, and f(x) extended-real-valued on $\mathbf{R}^n$.

For the sake of exposition we rewrite the inner optimization problem :

$$\rho(\omega,x) = \inf_{y \in Y(\omega,x)} \phi(\omega,y,x) \,. \tag{2.2}$$

Proper normal integrands (Rockafellar 1976 [121]) may be seen as extensions of proper convex normal integrands used above as well as extensions of the socalled Carathéodory integrands, to cover the case when functions attain infinite values. This concept received considerable attention for investigating *variational systems*.

Defining the multifunction epi $\phi(\omega,.,x)$ according to

$$\omega \rightarrow \text{epi}\, \phi(\omega,.,x) = \{(y,\alpha) \in \mathbf{R}^{n'} \times \mathbf{R} \mid \alpha \geq \phi(\omega,y,x)\} \tag{2.3}$$

we suppose now for every x in X :

(B1)' $\omega \to \text{epi}\ \phi(\omega,.,x)$ is a measurable, closed-valued multifunction ;
$y \to \phi(\omega,y,x)$ is proper for every $\omega \in \Omega$;
$\omega \to Y(\omega,x)$ is a measurable, closed-valued multifunction from Ω to $\mathbf{R}^{n'}$.

The first two conditions are precisely the definition of a proper normal integrand for $(\omega,y) \to \phi(\omega,y,x)$ ([121, p.173]). This and the third condition imply $\omega \to \rho(\omega,x)$ being measurable w.r.t. $(\Omega,\mathcal{B}^M)$ ([121, Theorem 2K]). For example, if $\phi(\omega,x,y)$ is finite, measurable in ω and continuous in y (i.e. Carathéodory integrand), for every x in X, the first two conditions in (B1)' are fulfilled ([121, Proposition 2C]). For illustrating the third condition of (B1)' we shall focus on those multifunctions $\omega \to Y(\omega,x)$ that allow to embed $\omega \to \rho(\omega,x)$ into a saddle function (below in section 6). For that purpose in particular, we consider the following parametric convex programming problem as counterpart to (1.6) :

$$\rho(\omega,x) := \inf\ q(\omega,y,x) \tag{2.4.1}$$
$$\text{s.t.}\ g_i^1(y,x) \leq h_i^1(\omega,x) \qquad i=1,...,r_1 \tag{2.4.2}$$
$$g_j^2(y,x) = h_j^2(\omega,x) \qquad j=1,...,r_2\ . \tag{2.4.3}$$

We assume that for every x in X :

(B1)'' - $y \to q(\omega,y,x)$ is proper convex, lower semicontinuous on $\mathbf{R}^{n'}$, for every $\omega \in \Omega$;
- $\omega \to q(\omega,y,x)$ is measurable w.r.t. $(\Omega,\mathcal{B}^M)$ for every $y \in \mathbf{R}^{n'}$;
- int $\{y \mid q(\omega,y,x) < +\infty\} \neq \emptyset$ for every $\omega \in \Omega$;
- $y \to g_i^1(y,x)$ is convex, real-valued on $\mathbf{R}^{n'}$ $(i=1,...,r_1)$;
- $y \to g_j^2(y,x)$ is linear on $\mathbf{R}^{n'}$ $(j=1,...,r_2)$;
- $\omega \to h_i^1(\omega,x)$ is measurable w.r.t $(\Omega,\mathcal{B}^M)$ $(i=1,...,r_1)$;
- $\omega \to h_j^2(\omega,x)$ is measurable w.r.t. $(\Omega,\mathcal{B}^M)$ $(j=1,...,r_2)$.

We realize that contrary to (A1)'' now also $q(\omega,y,x)$, $g_i^1(y,x)$ and $g_j^2(y,x)$ may depend on x as we allow nonconvexity of $x \to \rho(\omega,x)$. Hence for concluding the measurability of $\omega \to \rho(\omega,x)$ we may use the same argumentation as for (A1)''.

For measurability of optimal value functions consult also Kall & Oettli 1972 [76].

Remark 2.1 : In the above, assumptions for measurability has been discussed certainly not in the most general way, which would have been possible. For our purposes, we rather concentrated on special cases, easily available from the fundamental investigations in the above cited works of Rockafellar. (For a textbook on convex analysis and measurable multifunctions consult Castaing and Valadier 1977 [20]).

Continuing the interpretation for stochastic two-stage problems, we simply add: with (1.6),(2.4) one additionally allows to compensate the discrepancy to the targets $h_i^1(\omega,x)$, $h_j^2(\omega,x)$, likely occuring as soon as the outcome ω is known. This compensation is performed with the socalled optimal second-stage decision or recourse decision y, causing additional costs to be minimized; $q(\omega,y,x)$ represents the second-stage costs or recourse costs, whereby $\omega \rightarrow \rho(\omega,x)$ is called *recourse function* (implicitly thinking of the optimal value function of the underlying recourse problem).

Throughout this work we will speak of *stochastic (convex) two-stage problems*, also if $\omega \rightarrow \rho(\omega,x)$ is prescribed explicitly, as it has been supposed in (1.1) and (2.1). In these cases the second-stage costs are determined uniquely and might correspond to a fictitious second-stage decision. The essential assumption enabling such an interpretation is that we suppose to have full information of the outcome in the second stage. Thus, the feature that characterizes stochastic two-stage problems are less the 2 stages, but rather the associated 2 types of information (no vs. full) we suppose to have in these 2 stages.

Remark 2.2 : It might happen that full information becomes available due to successive increase in information over N(>2) stages. In this case it might be recommendable to use *multi-stage programs* for modeling. Such problems are in fact one of the challenging extensions mentioned in the beginning of this chapter.

Sofar, we have supposed that the extended-real-valued functions and probability measures defining (2.1) are precisely known. However, one has to take into account dynamic external or internal influences causing 'slight perturbations' of problem data. This fact raises the question whether the solutions of stochastic two-stage problems drastically change on slight perturbations or do they behave as 'slightly' (in the sense of continuous) as the perturbations. In other words we have to discuss stability of solution set and value. We make use of the results in Kall 1987 [71], Robinson and Wets 1987 [113] with the intention mainly to recall conditions and statements to illustrate the scope subject to which stability of optimal solutions and values is ensured.

In this section we concentrate on (2.1). We allow nonconvexity, restrict simultaneously on continuity of the involved functions, and additionally equip f(x) and $\rho(\omega,x)$ with a parameter denoted 'p', modeling the above-mentioned slight perturbations in accordance to Robinson and Wets 1987 [113]. In particular, we consider throughout this section the following parameterized version of (2.1) :

$$\inf_{x\in X} \ \psi(p,P,x) := f(p,x) + \int_{\Omega} \rho(p,\omega,x)\, dP(\omega) \tag{3.1}$$

with $p\in\Pi$, Π denoting a topological space. We adopt collections

$$\mathcal{F} := \{\rho(p,.,x) \mid p\in\Pi,\ x\in X\ ,\ \rho \text{ continuous on } \Pi\times\Omega\times X\ \}\ , \tag{3.2.1}$$
$$\mathcal{P} := \{P \mid P \text{ probability measure on } (\Omega,\mathcal{B}^M)\}, \tag{3.2.2}$$

being endowed with the topologies of uniform convergence on compact sets for $\mathcal{F}$ and weak convergence for $\mathcal{P}$. As illustrated in [113, p.1410] these topological spaces per se do not suffice to ensure continuity of the associated expectation operator $<\rho,P>=\int\rho dP$ w.r.t. $\mathcal{F}\times\mathcal{P}$. For that purpose a further restriction on $\mathcal{F}$ and $\mathcal{P}$ is needed; namely that of *uniform integrability* ([113, p.1410]):

Definition 3.1 : $\mathcal{F}$ is **uniformly integrable** w.r.t. $\mathcal{P}$ iff for each $\varepsilon>0$ there is a comp^ct set $\Omega(\varepsilon)$ in Ω such that for each $\rho\in\mathcal{F}$, each $P\in\mathcal{P}$ and $\Omega(\varepsilon)^c := \Omega \setminus \Omega(\varepsilon)$

$$\int_{\Omega(\varepsilon)^c} |\rho(p,\omega,x)|\, dP(\omega) < \varepsilon\ . \tag{3.3}$$

Therefore, *uniformly integrable* means that with respect to a certain tolerance level ε, it suffices to concentrate on those outcomes ω that lie in a certain ε-dependent compact set. Outcomes outside this compact set only have negligible impact.

Theorem 2.1. in [113] states that under the additional assumption of $\mathcal{F}$ uniformly integrable w.r.t. $\mathcal{P}$, $\langle\rho,P\rangle$ is continuous on $\mathcal{F}\times\mathcal{P}$. Herewith, having ensured continuity of the expectation operator, it remains to attack nonconvexity. This is done by employing the concepts of *epi-upper semicontinuity* and '*complete local minimizing set*' due to Robinson 1987 [111]. Epi-upper semicontinuity (with respect to p_0) for a parameterized function, say f(p,x), at x_0 is ensured e.g. for $f(.,x_0)$ being upper semicontinuous in p at p_0 (Robinson 1987 [111, Corollary 2.7]). Complete Local Minimizing sets (CLM) are characterized by the property that the nonempty set of local minimizers of an objective function subject to G (open, bounded) coincides with the set of local minimizers subject to clG; hence with respect to the closure of an open, bounded G the set cf local minimizers remains unchanged. Defining

$$\psi_G(p,P) := \inf \{ \psi(p,P,x) \mid x\in \mathrm{cl}G \} \qquad (3.4.1)$$

$$\Psi_G(p,P) := \{x\in \mathrm{cl}G \mid \psi(p,P,x) = \psi_G(p,P) \} \qquad (3.4.2)$$

we restate Theorem 2.2. in Robinson and Wets 1987 [113] as:

Proposition 3.1 : Let G be an open bounded set in $\mathbf{R}^n$, $\mathcal{P}$ be defined as in (3.2.2). Assuming the hypotheses: ρ is continuous on $\Pi\times\Omega\times\mathrm{cl}G$; the collection of functions $\mathcal{F}:= \{\rho(p,.,x) \mid p\in\Pi,\ x\in \mathrm{cl}G\}$ is uniformly integrable with respect to the family $\mathcal{P}$; the function f is lower semicontinuous on $\Pi\times\mathrm{cl}G$, and is epi-upper semicontinuous at some point $x_0 \in \Psi_G(p_0,P_0)$ with $f(p_0,x_0)$ finite; $\Psi_G(p_0,P_0)$ is a complete local minimizing set for $\psi(p_0,P_0,.)$ with respect to G. Then :

- $\psi_G(p_0,P_0)$ is finite and $\psi_G(p,P)$ is continuous at (p_0,P_0) ;
- $\Psi_G(p,P)$ is Berge-usc at (p_0,P_0) ;
- $\exists\ U\in \mathcal{N}(p_0,P_0)$ with $\forall\ (p,P)\in U$: $\psi(p,P,x)$ restricted to clG is proper and $\Psi_G(p,P)$ is a nonempty compact CLM set for $\psi(p,P,x)$ w.r.t. G .

Roughly speaking: under sufficiently 'slight' perturbations local minimizers remain complete and the optimal values behave continuously.

The results in Kall 1987 [71] are closely related to Theorem 2.2. in Robinson and Wets 1987 [113] for stochastic two-stage problems. In view to cover and to discuss also results of probabilistic constrained programs, the approach of [71] differs somewhat from that in [113].

Metricizing (at least locally) weak convergence on $\mathcal{P}$ stability results of quantitative type for mainly stochastic linear programming problems may be found (among other works) in Römisch and Walkolbinger 1987 [130], Römisch and Schultz 1991 [128-129]. In these works the authors succeeded to prove Lipschitz continuity for the optimal value and estimates for distances of optimal solutions. For a recent survey on stability and sensitivity results within stochastic programming consult Dupacová 1990 [37].

4 Epi-Convergence

Approximation schemes existing so far in literature within stochastic programming approximate successively either the recourse function ρ or the probability measure P. In both cases one solves the original problem through a sequence of approximate problems. To ensure that the approximate values and solutions get arbitrarily close to the optimal value and solution of the original problem, one has to employ a proper concept of convergence. In the last two decades the concept of epi-convergence has been developed, providing a unifying framework for approximating optimization problems. (Epi-convergence was introduced in the late 60s as a tool for approximating infinite dimensional problems in various fields of mathematics; references that correspond to early investigations of this theory may be found in Attouch and Wets 1981 [1].)

We review epi-convergence on the basis of countable many, extended-real-valued functions ψ, ψ^ν; ν=1,2,... , defined on the finite dimensional Euclidean space. (Relations of epi-convergence to the classical theory of convergence of functions have been worked out in Kall 1986 [69]).

Below, let Ψ, Ψ^ν (ν=1,2,...) denote the set of minimizers for ψ, ψ^ν; ν =1,2,... .

The characterization of epi-convergence and the subsequent proposition are recalled from Birge and Wets 1986 [14]. For the proof of Proposition 4.1 (stated below) see Attouch and Wets 1981 [1].

Definition 4.1 : Let $\{\psi, \psi^\nu; \nu=1,2,...\}$ be a collection of extended-real-valued functions defined on $\mathbf{R}^n$. The sequence $\{\psi^\nu; \nu=1,2,...\}$ is said to **epi-converge** to ψ (denoted $\psi^\nu \to_{epi} \psi$) iff for all $x \in \mathbf{R}^n$ we have :

$$\forall \{x^\nu; \nu=1,2,...\} \text{ converging to } x : \liminf_{\nu\to\infty} \psi^\nu(x^\nu) \geq \psi(x) \qquad (4.1.1)$$
$$\exists \{x^\nu; \nu=1,2,...\} \text{ converging to } x : \limsup_{\nu\to\infty} \psi^\nu(x^\nu) \leq \psi(x) . \qquad (4.1.2)$$

The epi-limit ψ is necessarily lower semicontinuous. The name epi-convergence stems from the fact that a collection of functions being epi-convergent to the epi-limit is equivalent to the convergence of the associated epi-graphs.

Proposition 4.1 : Suppose $\{\psi^\nu; \nu=1,2,...\}$ epi-converges to ψ . Then

$$\limsup_{\nu\to\infty} [\inf \psi^\nu] \leq \inf \psi \qquad (4.2.1)$$
$$\limsup_{\nu\to\infty} \Psi^\nu \subset \Psi \qquad (4.2.2)$$

Let $\{\psi^{\nu_k}, k=1,2,...\}$ denote a subsequence with $x^{\nu_k} \in \Psi^{\nu_k}$ and $x=\lim_{k\to\infty} x^{\nu_k}$. Then

$$\lim_{k\to\infty} [\inf \psi^{\nu_k}] = \inf \psi. \qquad (4.2.3)$$

If Ψ is nonempty, then

$$\lim_{\nu\to\infty} (\inf \psi^\nu) = \inf \psi \qquad (4.2.4)$$

iff $x \in \Psi$ implies the existence of sequences $\{\varepsilon^\nu \geq 0, \nu=1,2,...\}$, $\{x^\nu; \nu=1,2,...\}$ with $\lim_{\nu\to\infty} \varepsilon^\nu = 0$, $\lim_{\nu\to\infty} x^\nu = x$ and $x^\nu \in \varepsilon^\nu$-argmin ψ^ν.

Relation (4.2.1) states that all accumulation values of $\{\inf \psi^\nu; \nu=1,2,...\}$ are less or equal $\inf \psi$. (4.2.2) ensures that all accumulation points of the set of minimizers of ψ^ν are minimizers for ψ ; the associated limit of $\inf \psi^{\nu_k}$ exists and is equal to $\inf \psi$. However, example 4.1 (below) illustrates that the following situation can occur :

<u>Example 4.1</u> (Attouch and Wets 1981 [1]):

$$\psi^\nu(x) = \begin{cases} -1 & \text{if} \quad x < -\nu \\ \nu^{-1}\, x & \text{if} \quad -\nu \le x \le 0 \\ x & \text{if} \quad 0 < x \end{cases}$$

Obviously we have epi-convergence of ψ^ν to ψ, whereby $\psi(x):=x$ for x nonnegative and $\psi(x):=0$ otherwise); but, even with convexity, monotonicity, continuity and piecewise linearity we have: $\lim_{\nu\to\infty} [\inf \psi^\nu] = -1 < 0 = \inf \psi$ and $\Psi, \Psi^\nu \neq \emptyset$ for $\nu = 1,2,...$; this is due to the fact that $\limsup_{\nu\to\infty} \Psi^\nu = \emptyset$.

Hence, in the case no accumulation point exists (or at least some subsequence exists with no accumulation point) it might happen that

$$\lim_{\nu\to\infty} [\inf \psi^\nu] < \inf \psi \quad \text{(or at least : } \lim_{k\to\infty} [\inf \psi^{\nu_k}] < \inf \psi) . \qquad (4.3)$$

A necessary and sufficient condition that prohibits (4.3) is given with (4.2.4); its verification might become cumbersome due to the necessity that the sequence $\{\varepsilon^\nu \geq 0, \ \nu=1,2,...\}$ must be chosen appropriately. Things are getting a lot easier if there is a bounded set D with $D \cap \Psi^\nu \neq \emptyset$ for almost all ν. In this case any subsequence has some accumulation point entailing (4.2.4) - see Birge and Wets 1986 [14, Corollary 2.5]. A further sufficient condition entailing (4.2.4) for an epi-convergent sequence is available and stated as :

Proposition 4.2 : Suppose $\{\psi^\nu; \nu=1,2,...\}$ epi-converges to ψ, $\limsup_{\nu\to\infty} \Psi^\nu \neq \emptyset$ and $\{\inf \psi^\nu; \nu = 1,2,...\}$ are monotone increasing or monotone decreasing. Then, the infima converge and $\lim_{\nu\to\infty} (\inf \psi^\nu) = \inf \psi$.

Proof: The assertion follows immediately from $\inf \psi^\nu$ being monotone and the fact that a convergent subsequence of minimizers exists, implying (4.2.3). //

Summarizing we may state that the verification of (4.2.4) becomes difficult if some subsequence $\{\psi^{\nu_k}\}$ exists whose minimizers do not accumulate and no bounded set covering at least some minimizers of almost all approximates, no monotonicity of the inf-values respectively, is at hand.

We next recall useful criteria for verifying epi-convergence of a sequence of functions (see Birge and Wets 1986 [14, pp 58-62]):

Proposition 4.3 : Suppose ψ is lower semicontinuous on $\mathbf{R}^n$. Let $\{\psi^\nu;\ \nu=1,2,...\}$ be a collection of functions converging pointwise to ψ. Then $\{\psi^\nu;\ \nu=1,2,...\}$ epi-converges to ψ, if they are monotone increasing or monotone decreasing.

Proposition 4.4 : Suppose $\{P^\nu,\ \nu=1,2,...\}$ is a sequence of probability measures converging weakly to P defined on $(\Omega,\mathcal{B}^M)$. Let $\rho:\Omega\times\mathbf{R}^n\rightarrow\mathbf{R}\cup\{\infty\}$ be continuous in ω for every x in dom $E\rho := \{x \mid (E\rho)(x)<\infty\}$, and be locally Lipschitz in x on dom $E\rho$ with Lipschitz constant independent of ω; further, $\{\rho(.,x) \mid x\in \text{dom } E\rho\}$ are assumed to be uniformly integrable with respect to $\{P,\ P^\nu\ (\nu=1,2,...)\}$ and with $P\{\omega|\rho(\omega,x)=\infty\}>0$ if and only if $P^\nu\{\omega|\rho(\omega,x)=\infty\}>0$ for almost all ν. Let $(E^\nu\rho)(x)$ denote the expectation functionals of $\rho(.,x)$ with respect to P^ν. Then $\{E^\nu\rho\ ;\ \nu=1,2,...\}$ epi- and pointwise converges to $E\rho$.

Above propositions will become essential for the results in chapter III. Therein, we shall construct a sequence of probability measures $\{P,\ P^\nu\ (\nu=1,2,...)\}$ that converge weakly to P, and additionally entail monotonicity of the associated expectation functionals $(E^\nu\rho)(x)$ $(\nu=1,2,...)$. This, together with pointwise convergence of $\{(E^\nu\rho)(x);\ \nu=1,2,...\}$ allow to conclude epi-convergence of $\{(E^\nu\rho)(x);\ \nu=1,2,...\}$ in the case $(E\rho)(x)$ being lower semicontinuous.

5 *Saddle Property*

Throughout this section, we aim at structural properties of the recourse function $\omega\rightarrow\rho(\omega,x)$. In particular, we shall imbed $\omega\rightarrow\rho(\omega,x)$ in a saddle function, if $\rho(.,x)$ represents the optimal value function of an 'inner' convex optimization problem of type (1.6) or (2.4). Let us focus on (2.4) being restated for the ease of exposition :

$$\rho(\omega,x) := \inf\ q(\omega,y,x) \tag{5.1.1}$$

$$\text{s.t.}\quad g_i^1(y,x) \le h_i^1(\omega,x) \qquad i=1,...,r_1 \tag{5.1.2}$$

$$g_j^2(y,x) = h_j^2(\omega,x) \qquad j=1,...,r_2 \tag{5.1.3}$$

We suppose assumption (B1)" to hold. Due to the results of convex analysis we may state the following well-known properties, adapted to our context (see e.g. Kall and Stoyan 1982 [79, Lemma 3]) :

Proposition 5.1 (is restated more generally in chapter II as Proposition II.1):

For every x in X :
i) If $q(.,.,x)$ is deterministic (i.e. $q(.,y,x)$ is constant on Ω for any y), and if $h_i^1(.,x)$ $(i=1,...,r_1)$, $h_j^2(.,x)$ $(j=1,...,r_2)$ are linear affine in ω, then the recourse function $\omega \rightarrow \rho(\omega,x)$ is **convex**.
ii) If right-hand sides in (5.1.2), (5.1.3) are deterministic (i.e. $h_i^1(.,x)$ $(i=1,...,r_1)$, $h_j^2(.,x)$ $(j=1,...,r_2)$ are constant on Ω), and if $\omega \rightarrow q(\omega,y,x)$ is concave in ω on Ω for every y, then the recourse function $\omega \rightarrow \rho(\omega,x)$ is **concave**.

We will illustrate that above features allow to deduce a recourse function, denoted $\rho^s(.,.,x)$, that is concave-convex (for every x in X) and may substitute for $\rho(\omega,x)$ in the associated stochastic program. Due to the additional assumptions stated in i) , ii) of Proposition 5.1, we have to strengthen (B1)" for every x in X :

(B1)"' - $y \rightarrow q(\omega,y,x)$ is proper convex, lower semicontinuous on $\mathbf{R}^{n'}$, for every $\omega \in \Omega$;
- $\omega \rightarrow q(\omega,y,x)$ is measurable w.r.t. $(\Omega,\mathcal{B}^M)$, **concave** for every $y \in \mathbf{R}^{n'}$;
- int $\{y \mid q(\omega,y,x) < +\infty\} \neq \emptyset$ for every $\omega \in \Omega$;
- $y \rightarrow g_i^1(y,x)$ is convex, real-valued on $\mathbf{R}^{n'}$ $(i=1,...,r_1)$;
- $y \rightarrow g_j^2(y,x)$ is linear on $\mathbf{R}^{n'}$ $(j=1,...,r_2)$;
- $\omega \rightarrow h_i^1(\omega,x)$ is measurable w.r.t. $(\Omega,\mathcal{B}^M)$, **linear affine** $(i=1,...,r_1)$;
- $\omega \rightarrow h_j^2(\omega,x)$ is measurable w.r.t. $(\Omega,\mathcal{B}^M)$, **linear affine** $(j=1,...,r_2)$.

Let pr^o, pr^r denote projections of ω into $\mathbf{R}^K$, $\mathbf{R}^L$, i.e.

$$pr^o : \mathbf{R}^M \rightarrow \mathbf{R}^K \quad , \qquad (5.2.1)$$
$$pr^r : \mathbf{R}^M \rightarrow \mathbf{R}^L \quad , \qquad (5.2.2)$$

with the property that $pr^o(\omega)$ represents that subvector of ω which influences the objective function $q(\omega,.,x)$ significantly, in the sense that $q(.,y,x)$ is <u>not constant</u> with respect to any component of $pr^o(\omega)$ for any y and any x in X. Similarly $pr^r(\omega)$ is assumed to represent that subvector of ω that influences the right-hand side of the constraints (5.1.2),(5.1.3) significantly, in the sense that $h_i^1(.,x)$

$(i=1,\dots,r_1)$, $h_j^2(.,x)$ $(j=1,\dots,r_2)$ are not constant subject to any component of $pr^r(\omega)$, for any x in X.

Remark 5.1.[1] : We have assumed that ω is the random vector with known probability measure P. It might happen that there exist components which influence neither the right-hand side nor the objective. Inhowfar we are able to reduce the dimension to those components that influence at least one part depends on the ability of determining the associated marginal distribution of P.
In the case that all components of ω influence the right-hand side as well as the objective, then K=L=M and pr^o, pr^r are the identity mapping.

Defining

$$\Omega^o := \text{cl}\,[\text{co}\{\ \eta \mid \eta=pr^o(\omega),\ \omega\in\Omega\ \}] \tag{5.3.1}$$

$$\Omega^r := \text{cl}\,[\text{co}\{\ \xi \mid \xi=pr^r(\omega),\ \omega\in\Omega\ \}] \tag{5.3.2}$$

we may introduce a new collection of parametric convex mathematical programs for every x in X :

$$\rho^s(\eta,\xi,x) := \inf\ q(\eta,y,x) \tag{5.4.1}$$

$$\text{s.t.}\ \ g_i^1(y,x) \le h_i^1(\xi,x) \qquad i=1,\dots,r_1 \tag{5.4.2}$$

$$g_j^2(y,x) = h_j^2(\xi,x) \qquad j=1,\dots,r_2\ . \tag{5.4.3}$$

Observe, that $h_i^1(.,x)$ and $h_j^2(.,x)$ are now defined on $\Omega^r\subset\mathbf{R}^L$ instead of $\Omega\subset\mathbf{R}^M$ as in (5.1). Analogously $q(.,y,x)$ is now given on $\Omega^o\subset\mathbf{R}^K$ instead of $\Omega\subset\mathbf{R}^M$. Due to the properties of pr^o, pr^r this causes no essential difference to the situation in (5.1). There is no loss of information if we ignore those subspaces subject to which $q(.,y,x)$, $h_i^1(.,x)$, $h_j^2(.,x)$ are constant. Nevertheless, we have to pay attention to the measurability of the optimal value function $\rho^s(.,.,x)$ with respect to $\mathcal{B}^{K+L}=\mathcal{B}^K\times\mathcal{B}^L$, where $\mathcal{B}^{K+L},\mathcal{B}^K,\mathcal{B}^L$ denote the Borel sets restricted to $\Omega^o\times\Omega^r,\Omega^o,\Omega^r$ ($\mathcal{B}^{K+L}=\mathcal{B}^K\times\mathcal{B}^L$ holds due to separability of the Euclidean space, see e.g. Billingsley 1968 [9, p.225]). For the ease of exposition we formally outline the measurability conditions for $\rho^s(\eta,\xi,x)$ being supposed to hold and seen as counterparts to (B1)''' for every x in X - imposing no additional restriction:

---

1) I owe thanks to Prof. S.M. Robinson for pointing out the incompleteness of the discussion in a previous version.

(B1)^s - $y \to q(\eta,y,x)$ is proper convex, lower semicontinuous on $\mathbf{R}^{n'}$, for every $\eta \in \Omega^o$;
- $\eta \to q(\eta,y,x)$ is measurable w.r.t. $(\Omega^o, \mathcal{B}^K)$, **concave** for every $y \in \mathbf{R}^{n'}$;
- int $\{y \mid q(\eta,y,x) < +\infty\} \neq \emptyset$ for every $\eta \in \Omega^o$;
- $y \to g_i^1(y,x)$ is convex, real-valued on $\mathbf{R}^{n'}$ $(i=1,\ldots,r_1)$;
- $y \to g_j^2(y,x)$ is linear on $\mathbf{R}^{n'}$ $(j=1,\ldots,r_2)$;
- $\xi \to h_i^1(\xi,x)$ is measurable w.r.t. $(\Omega^r, \mathcal{B}^L)$, **linear affine** $(i=1,\ldots,r_1)$;
- $\xi \to h_j^2(\xi,x)$ is measurable w.r.t. $(\Omega^r, \mathcal{B}^L)$, **linear affine** $(j=1,\ldots,r_2)$.

Remark 5.2 : It is obvious that in the case we work with convex programs (1.6) instead of (2.4) we have to strengthen assumption (A1)" correspondingly.

Next, having a measurable transformation $\chi : \Omega \to \Omega^o \times \Omega^r$ given according to

$$\chi(\omega) = pr^o(\omega) \times pr^r(\omega) \qquad \text{for all } \omega \in \Omega \tag{5.5}$$

it is easy to characterize that probability measure on $(\Omega^o \times \Omega^r, \mathcal{B}^{K+L})$ which is induced by P on $(\Omega, \mathcal{B}^M)$ through $\chi(\omega)$,

$$P^\chi(B^{K+L}) := P(\chi^{-1}(B^{K+L})) \text{for all } B^{K+L} \in \mathcal{B}^{K+L} \tag{5.6}$$

We obviously have $\rho^s(\chi(\omega),x) = \rho(\omega,x)$ for all $\omega \in \Omega$. Applying the transformation theorem (see e.g. Halmos 1974 [55, Theorem C, p.163]) yields

$$\int_{\Omega^o \times \Omega^r} \rho^s(\eta,\xi,x) dP^\chi(\eta,\xi) = \int_\Omega \rho^s(\chi(\omega),x) dP(\omega) = \int_\Omega \rho(\omega,x) dP(\omega). \tag{5.7}$$

Hence, supposing some (at least 'hidden') concavity, convexity of $\rho(.,x)$ in the random components, as is the case within recourse problems of the type (1.6) or (2.4), we rewrite stochastic programs (1.1), (1.2) respectively, equivalently as:

$$\inf_{x \in X} \quad f(x) + \int_{\Omega^o \times \Omega^r} \rho^s(\eta,\xi,x) dP^\chi(\eta,\xi) \, . \tag{5.8}$$

Now $\rho^s(.,\xi,x)$ is concave in η (on Ω^o) for all $\xi \in \Omega^r$, $\rho^s(\eta,.,x)$ is convex in ξ (on Ω^r) for all $\eta \in \Omega^o$, for every x in X. The price we have to pay for raising the saddle property on the recourse function is an increase in the dimension of the probability space. If P is discrete (either some empirical distribution or some sample distribution) this actually is no loss; we simply perform the transformation χ on these atoms to obtain the induced discrete probability measure P^χ. On the other hand, we have to stress that, if P is absolutely continuous, this need not be inherited by P^{χ} (as the example 5.1 below illustrates). Thus, is the saddle property preferable to absolute continuity ? This question is a challenge per se and requires thorough investigations of computational behaviour in dependence on different structural properties of stochastic two-stage programs. We postpone such investigations to a future work.

The impact of working with saddle functions $\rho^s(.,.,x)$ becomes apparent in sections 14 and the following. To illustrate the above technique for raising the saddle property, we consider the following convex program, allowing an explicit representation of the optimal value function:

Example 5.1: We suppose $X=\mathbf{R}$, $y\in\mathbf{R}^2$, $\Omega=co[(0,1),(1,0),(0,-1)]\subset\mathbf{R}^2$ be a simplex; P is the uniform distribution on Ω; i.e. $\omega=(\omega_1,\omega_2)$ uniformly distributed in Ω. The convex program is given with

$$\inf\{\omega_1.y_1^2+\omega_1.y_2 \mid y_1=\omega_1\,,\ y_2=\omega_2.x\} \tag{5.9}$$

yielding the optimal value function $\rho(\omega,x)=\omega_1^3+\omega_1.\omega_2.x$. Due to the above we set

$$pr^o(\omega_1,\omega_2)=\omega_1=\eta_1 \tag{5.10.1}$$
$$pr^r(\omega_1,\omega_2)=(\omega_1,\omega_2)=(\xi_1,\xi_2)\ . \tag{5.10.2}$$

This yields the new parametric convex program

$$\inf\{\eta_1.y_1^2+\eta_1.y_2 \mid y_1=\xi_1\,,\ y_2=\xi_2.x\} \tag{5.11}$$

for $\eta_1\in\Omega^o=[0,1]$ and $\xi\in\Omega^r=co\{(0,1),(1,0),(0,-1)\}$ with the associated optimal value function $\rho^s(\eta,\xi,x)=\eta_1.\xi_1^2+\eta_1.\xi_2.x$ being a saddle function in (η_1,ξ_1,ξ_2) on $\Omega^o\times\Omega^r$. Setting

$$\chi(\omega_1,\omega_2)=(\omega_1,\omega_1,\omega_2)=(\eta_1,\xi_1,\xi_2)\ . \tag{5.12}$$

we obtain that (η_1,ξ_1,ξ_2) is uniformly distributed on $co\{(0,0,1),(1,1,0),(0,0,-1)\}$, a 2-dimensional simplex in $\mathbf{R}^3$; i.e. P^{χ} represents the uniform distribution on $co\{(0,0,1),(1,1,0),(0,0\text{-}1)\}$, and is therefore not absolutely continuous.

Below we suppose to have stochastic independent subvectors within the random data ω. Say, we have with $\omega=(\omega^1,\ldots,\omega^I)$ I random vectors ω^i $(i=1,\ldots,I)$ that are independent and are distributed according to some probability measures P^i defined on $(\Omega^i, \mathcal{B}^{M_i})$. For motivating the following, let the probability measures P^i be available as empirical distributions with support Ω^i containing 10,000 data for each $i=1,\ldots,I$. For $I=3$ and $M_i=1$ for $i=1,2,3$ this amounts to work with 30,000 data, being certainly manageable. On the other hand, without having the possibility of exploiting stochastic independence this would amount to work with 10^{12} data that might become unaffordable. As a consequence one would have to rely on a sample information causing additional statistical error, which could have been avoided.

Therefore, intending to design approximation schemes for stochastic programs (5.8) we certainly are better with focusing on probability spaces $(\Omega^{M_i},\mathcal{B}^{M_i},P^i)$ separately for $i=1,\ldots,I$. Similarly, aiming at the saddle property of the recourse function, we apply the technique of the preceding section separately on ω^i, $i=1,\ldots,I$. For completeness we do that below.

With

$$(\Omega,\mathcal{B}^M,P) = \mathop{\mathsf{X}}_{i=1}^{I} (\Omega^i, \mathcal{B}^{M_i},P^i) , \tag{6.1}$$

the stochastic program reads as

$$\inf_{x\in X} \quad f(x) + \int_{\Omega^1} \ldots \int_{\Omega^I} \rho(\omega^1,\ldots,\omega^I,x) \, dP^I(\omega^I)\ldots dP^1(\omega^1) \quad . \tag{6.2}$$

We apply (5.2), (5.5), (5.6) to ω^i for all $i=1,\ldots,I$ and get self-explaining in notation (according to the preceding section) :

$$pr^{o,i}(\omega^i) : \Omega^i \to \mathbf{R}^{K_i} \tag{6.3.1}$$
$$pr^{r,i}(\omega^i) : \Omega^i \to \mathbf{R}^{L_i} \tag{6.3.2}$$
$$\Omega^{o,i} := cl[co\{ \eta^i \mid \eta^i = pr^{o,i}(\omega^i) , \omega^i \in \Omega^i \}] \tag{6.3.3}$$
$$\Omega^{r,i} := cl[co\{ \xi^i \mid \xi^i = pr^{r,i}(\omega^i) , \omega^i \in \Omega^i \}] \tag{6.3.4}$$
$$\chi^i(\omega^i) = pr^{o,i}(\omega^i) \times pr^{r,i}(\omega^i) \tag{6.3.5}$$
$$P^{\chi,i}[B^{K_i+L_i}] := P[(\chi^i)^{-1}(B^{K_i+L_i})] \qquad \forall\, B^{K_i+L_i} \in \mathcal{B}^{K_i+L_i} . \tag{6.3.6}$$

With setting

$$\eta := \times_i \eta^i := \times_i \mathrm{pr}^{o,i}(\omega^i) \tag{6.4.1}$$

$$\xi := \times_i \xi^i := \times_i \mathrm{pr}^{r,i}(\omega^i) \tag{6.4.2}$$

and with $(B1)^s$ of section 5 supposed to hold, we obtain $\rho^s(\eta,\xi,x)$ measurable on $\times_i(\Omega^{o,i}\times\Omega^{r,i},\mathcal{B}^{K_i+L_i})$ for every x in X. Having χ: $\times_i \Omega^{o,i} \to \times_i (\Omega^{o,i}\times\Omega^{r,i})$

$$\chi(\omega) := (\chi^1(\omega^1), \ldots, \chi^I(\omega^I)) \tag{6.5}$$

as measurable transformation we get as counterpart to (5.7) :

$$\int_{\Omega^{o,1}\times\Omega^{r,1}} \cdots \int_{\Omega^{o,I}\times\Omega^{r,I}} \rho^s(\eta^1,\xi^1,\ldots,\eta^I,\xi^I,x)\, dP^{\chi,I}(\eta^I,\xi^I)\ldots dP^{\chi,1}(\eta^1,\xi^1) =$$

$$= \int_{\Omega^1} \cdots \int_{\Omega^I} \rho^s(\chi^1(\omega^1),\ldots,\chi^I(\omega^I),x)\, dP^I(\omega^I)\ldots dP^1(\omega^1) = \tag{6.6}$$

$$= \int_{\Omega^1} \cdots \int_{\Omega^I} \rho(\omega^1,\ldots,\omega^I,x)\, dP^I(\omega^I)\ldots dP^1(\omega^1) .$$

Herewith, we write the stochastic program (6.2) equivalently as

$$\inf_{x\in X} f(x) + \int_{\Omega^{o,1}\times\Omega^{r,1}} \cdots \int_{\Omega^{o,I}\times\Omega^{r,I}} \rho^s(\eta^1,\xi^1,\ldots,\eta^I,\xi^I,x)\, dP^{\chi,I}(\eta^I,\xi^I)\ldots dP^{\chi,1}(\eta^1,\xi^1) \tag{6.7}$$

with now $\rho^s(\eta^1,\xi^1,\ldots,\eta^I,\xi^I,x)$ being concave in $(\eta^1,\ldots,\eta^I)$ on $\Omega^o := \times_i \Omega^{o,i}$ for every $\xi \in \Omega^r := \times_i \Omega^{r,i}$, and being convex in $(\xi^1,\ldots,\xi^I)$ on Ω^r for every $\eta \in \Omega^o$, $\forall$ $x \in X$.

Modelling stochastic independence within raising the saddle property is rather obvious; in which way we *preserve* stochastic independence throughout approximations designed to solve (6.7) will become apparent in chapter III.
From now on, we will focus on stochastic programs being given as in (5.8), or as in (6.7), with omitting s and χ as superscripts of ρ and P as there will be no danger of confusion.

In this section we illustrate special convex cases of stochastic two-stage models, well-established in literature. For the ease of exposition we restrict on (I=1):

$$\inf_{x\in X} \; f(x) + \int_{\Omega^o\times\Omega^r} \rho(\eta,\xi,x)dP(\eta,\xi) \;. \qquad (7.1)$$

The linear version of (7.1) is the well-known *stochastic linear program with recourse* and is given as

$$\inf \; \langle c,x\rangle + \int_{\Omega^o\times\Omega^r} \rho(\eta,\xi,x)dP(\eta,\xi) \qquad (7.2.1)$$

$$\text{s.t.} \quad Ax \le a$$

$$x \ge 0$$

with

$$\rho(\eta,\xi,x) := \inf \; \langle q(\eta),y\rangle \qquad (7.2.2)$$

$$\text{s.t.:} \quad Wy=h(\xi)-T(\xi)x$$

$$y\ge 0 \;.$$

$X=\{x|\; Ax \le a,\; x \ge 0\}$ represents a convex polyhedral set of $\mathbf{R}^n$ given by the matrix $A\in\mathbf{R}^{m\times n}$ and vector $a\in\mathbf{R}^m$; problem (7.2.2) is called *linear recourse problem* serving to compensate $h(\xi)-T(\xi)x$ under minimal additional costs. $W\in\mathbf{R}^{m'\times n'}$ represents the socalled *fixed* (in the sense of deterministic) recourse matrix. If W is of full rank with respect to the nonnegative orthants (i.e. $\{Wy|y\ge 0\}=\mathbf{R}^{m'}$), we speak of *complete recourse* allowing compensation for all $x\in\mathbf{R}^n$; *relatively complete recourse* refers to the case that for any x in X and any ξ in Ω^r there always exists some $y\ge 0$ so that $Wy=h(\xi)-T(\xi)x$. Both entail that the recourse function is less than $+\infty$ on Ω^r for any feasible x. If we have additionally dual feasibility P-almost surely in (7.2.2), i.e. $\{u|W'u\le q(\eta)\}\ne\emptyset$ P-a.s., then the recourse function is finite (P-a.s.) for any x in X. Integrability of the recourse function (w.r.t. P) is ensured if the cross moments of $q(\eta),h(\xi),T(\xi)$ exist with respect to P. Further, q,h,T are usually assumed to be linear affine in η,ξ, i.e.:

$$q(\eta) = q_0 + \sum_{k=1}^{K} q_k.\eta_k \,,\quad h(\xi) = h_0 + \sum_{l=1}^{L} h_l\,.\xi_l \,,\quad T(\xi) = T_0 + \sum_{l=1}^{L} T_l\,.\xi_l \,, \qquad (7.3)$$

with $q_k\in\mathbf{R}^{n'}$, $h_l\in\mathbf{R}^{m'}$, $T_l\in\mathbf{R}^{m'\times n}$ (k=0,...,K ; l=0,...,L).

Remark 7.1 : The motivation for (7.3) is to reduce the dimension of the random vector if some uncertain problem data depend linearly on random outcomes (see Kall 1976 [66]). In the sequel we often distinguish cases in which $q(\eta)$, $h(\xi)$ or $T(\xi)$ are assumed to be deterministic. In these cases we commonly substitute $q:=q(\eta)$, $h:=h(\xi)$ or $T:=T(\xi)$. Further, in case that ξ influences only h (the right-hand side), then one usually allows that ξ may be taken of the same dimension as h; we indicate this situation with setting $\xi := h(\xi)$.

Applying Proposition 5.1 to stochastic linear two-stage problem (7.2) yields:

Proposition 7.1: i) $\forall\ x\in X\ \ \rho(.,.,x)\colon \Omega^o\times\Omega^r\to\mathbf{R}$ are concave-convex piecewise linear saddle-functions; ii) $\forall\ (\eta,\xi)\in\Omega^o\times\Omega^r\ \ \rho(\eta,\xi,.)\colon X\to\mathbf{R}$ are convex piecewise linear functions.

Trivially, the recourse function $\rho(\eta,\xi,x)$ reduces to a convex function in ξ,x separately, if the second-stage costs are deterministic (i.e. $q(\eta)=q$); moreover, we have convexity in (ξ,x) if the technology matrix is deterministic ($T(\xi)=T$), too. On the other hand $\rho(\eta,\xi,x)$ reduces to a saddle function in (η,x) if both h and T are fixed ($h(\xi)=h$, $T(\xi)=T$). In all these cases, piecewise linearity in the random arguments and in x still holds.

A common case subject to which things become more comfortable is the case of *simple recourse* (i.e. $W=(I,-I)$) with deterministic T and q. Substituting $(y^+,y^-):=y$, $(q^+,q^-):=q$, $h(\xi)=\xi\in\mathbf{R}^{m'}$ within (7.2.2) yields the following recourse function:

$$\rho(\xi,x) := \sum_{j=1}^{m'} \rho_j(\xi_j ,(Tx)_j) \tag{7.4.1}$$

with

$$\begin{aligned} \rho_j(\xi_j ,(Tx)_j) := \ & \inf\ q_j^+ y_j^+ + q_j^- y_j^- \\ \text{subject to}\quad & y_j^+ - y_j^- = \xi_j - (Tx)_j \\ & y_j^+, y_j^- \geq 0 . \end{aligned} \tag{7.4.2}$$

Due to the gained separability of the recourse function in the components of ξ, it suffices to consider m' one-dimensional marginal distributions of P, denoted with P_j (j=1,...,m') and corresponding support Ω_j^r. We then get for the objective of the underlying problem with simple recourse:

$$\inf_{x\in X} \; c'x + \sum_{j=1}^{m'} \int_{\Omega_j^r} \rho_j(\xi_j, (Tx)_j)\, dP_j(\xi_j) \,. \tag{7.5}$$

For the results stated above and further thorough investigations of stochastic linear programs with recourse, consult Wets 1974 [141] and Kall 1976 [66]. For a recent text book on this topic see Klein Haneveld 1986 [82] .

Allowing f(x) in (7.1) to be quadratic (convex), technology matrix $T(\xi)$, right-hand side vector $\xi=h(\xi)$ to be stochastic and introducing a piecewise linear-quadratic function in $[\xi_j-(Tx)_j]$ as stochastic penalty function in (7.4), one obtains the socalled *stochastic quadratic program with simple recourse* proposed in Rockafellar and Wets 1986 [124]. A generalization of this model called *stochastic linear-quadratic programming problem* is presented in Rockafellar and Wets 1986 [125] and is stated herein (slightly modified with respect to our context) as:

$$\inf_{x\in X} \quad \langle c,x\rangle + \langle x,Cx\rangle + \int_{\Omega^o\times\Omega^r} \rho(\eta,\xi,x)dP(\eta,\xi) \tag{7.6.1}$$

with

$$\rho(\eta,\xi,x) := \sup \{\langle y,[h(\xi)-T(\xi)x]\rangle - 1/2\langle y,H(\xi)y\rangle\} \quad \text{s.t.:} \quad By = b(\eta) \,. \tag{7.6.2}$$

X denotes a convex polyhedral set; $h(\xi)\in \mathbf{R}^{n'}$, $T(\xi)\in \mathbf{R}^{n'\times n}$, $H(\xi)\in \mathbf{R}^{n'\times n'}$, $b(\eta)\in \mathbf{R}^{m'}$, are linear affine in ξ and η similarly due to (7.3); further C and $H(\xi)$ (symmetric) supposed to be positive definite for all ξ in Ω^r and $B\in \mathbf{R}^{m'\times n'}$. Thus, the stochastic linear-quadratic model covers the case when penalty expressions are introduced to restrain the difference $h(\xi)-T(\xi)x$ [125, p.64]. We easily verify that herewith $\rho(.,.,x)$ is a saddle function for any x in X.
The necessity to consider measurability under convexity on some subset becomes apparent due to the above: it only makes sense to suppose $H(\xi)$ being positive definite with respect to an appropriate subset Ω^r, instead of the entire $\mathbf{R}^K$.

Last but not least we have to mention the important case when the underlying probability space is discrete with finite support $\Omega=\{(\eta,\xi)^1,\ldots,(\eta,\xi)^N\}$ with $P[(\eta,\xi)^j]=p_j$, j=1,...,N. If N is not too large, discreteness allows to rewrite (7.1) as

$$\inf_{x\in X} \; f(x) + \sum_{j=1}^{N} \rho((\eta,\xi)^j,x).p_j \tag{7.7}$$

and consider the corresponding deterministic equivalent problem. If $\rho(\eta,\xi,x)$ is given as optimal value function of a parametric convex program, this amounts to solve N convex mathematical programs, more precisely: we have to minimize with respect to y subject to given x and $(\eta,\xi)^j$ $(j=1,\dots,N)$. Composing both minimizations (the inner minimization w.r.t. y and the outer w.r.t. x) to one minimization w.r.t $(x,y^1,\dots,y^N)$, we end with a block-structured convex program of the form :

$$\inf \quad f(x) + q(\eta^1,y^1).p_1 + \dots + q(\eta^N,y^N).p_N \tag{7.8.1}$$

$$\text{s.t.:} \quad -h_i^1(\xi^1,x) + g_i^1(y^1) \leq 0 \quad i=1,\dots,r_1 \tag{7.8.2}$$

$$-h_j^2(\xi^1,x) + g_j^2(y^1) = 0 \quad j=1,\dots,r_2 \tag{7.8.3}$$

$$\dots \qquad \dots$$

$$-h_i^1(\xi^N,x) \qquad +g_i^1(y^N) \leq 0 \quad i=1,\dots,r_1 \tag{7.8.4}$$

$$-h_j^2(\xi^N,x) \qquad +g_j^2(y^N) = 0 \quad j=1,\dots,r_2 \tag{7.8.5}$$

$$x \in X\ ,\ y^1,\dots,y^N \geq 0\ .$$

Besides dual block-angularity of program (7.8) we have to stress that $g_i^1(y^1)$, $g_j^2(y^1)$ $(i=1,\dots,r_1\ ,\ j=1,\dots,r_2)$ remain unchanged throughout the N blocks. For solving such programs it is strongly recommended to employ methods exploiting both, block structure as well as unchanged data among the blocks, so that the computational effort remains affordable with increasing N. Up to now there exists a number of methods which have been designed especially for exploiting these features: For the linear case we mention i) the L-Shaped Method due to Van Slyke & Wets 1969 [140] and its extensions due to Birge 1985 [10], 1988 [12], and Gassman 1990 [52]; ii) the Basis-Factorization Method due to Strazicky 1980 [137], Kall 1979 [67] and iii) the Regularized Decomposition Method due to Ruszczynski 1986 [132] : For a survey on large-scale linear techniques we refer to Wets 1988 [143]. For the linear-quadratic case we have the Lagrangian Finite Generation Method due to Rockafellar and Wets 1986 [125], the Finite Envelope Method due to Rockafellar 1990 [122] and the Progressive Hedging Algorithm due to Rockafellar and Wets 1991 [127]. Most of these methods are still under investigation for further improvements and extensions, considering also parallelizing aspects for the usage of parallel or vector computers.

Remark 7.2: Considering finite discrete probability measures in stochastic programs of type (2.1) we similarly obtain a block-structured nonconvex program with the above-mentioned features. Sofar, we can only refer to the Progressive Hedging Algorithm that might allow to be extended in that direction according to the discussion of Rockafellar and Wets 1991 [127, pp. 145-146].

Sofar, we aimed at motivating and specifying (convex) stochastic two-stage problems (1.1), (2.1). We concentrated on assumptions ensuring well-posedness, stability and convergence of approximates. We further raised the saddle property for the recourse function with simultaneously preserving stochastic independence. Subsequently (chapter II) we focus on (semi-)continuity and subdifferentiability properties of optimal value functions corresponding to parameterized convex programs of type (5.4). This necessitates to investigate duality relations for the parametric convex program. These results might be viewed as extension of the corresponding one for convex optimization problems convexly parameterized. In chapter III we shall use these duality results for analysing the goodness of the achieved approximates.

II Duality and Stability in Convex Optimization (Extended Results for the Saddle Case)

Due to the first chapter we realize that convex programs of the type (5.4) with random parameters (η,ξ) in the objective and in the right-hand side play an essential role in stochastic two-stage programs. We are aware of the fact that we have to integrate (at least approximately) the optimal value function (recourse function) of (5.4), given - for a fixed first-stage decision x - only implicitly by the infimum of (5.4). Although the optimal value function changes with x, we take into account that it remains a saddle function. This fact motivates to derive suitable approximates for the optimal value function, that are computable and allow conclusions about the goodness of the first-stage decision x. Such approximates will be derived in chapter III. At this point it is only of importance to stress that the existence of these approximates essentially depend on the existence of subgradients of the optimal value function at distinguished points. Hence, at first, we are confronted with investigating under which assumptions subgradients for the optimal value function at a certain distinguished point, say (η^0,ξ^0), exist. In other words we have to ensure dual solvability of the convex optimization problem - of the type (5.4) - with respect to (η,ξ) at (η^0,ξ^0).

The above turns us to the problem of studying subdifferentiability and continuity of optimal value functions at (η^0,ξ^0), that correspond to (finite dimensional) convex optimization problems with (η,ξ) as parameters. This well known task (within *duality theory*) is a challenge per se and independent of any stochastic model in the background. Throughout this chapter we therefore free us from notions like 'first second stage decision', 'recourse function' etc. Instead, for the ease of understanding, we will study subdifferentiability and continuity of optimal value functions with an autonomous notation: in particular, subsequent investigations base on denoting with $f(\eta,\xi,x)$ the parametric objective function of a primal convex optimization problem; $f(\eta,\xi,x)$ is intended to be minimized with respect to x, that might represent below e.g. a decision, production plan etc. ; the associated (primal) optimal value function will be denoted $\phi(\eta,\xi)$.

First, let us begin with some economical interpretation, that motivates to study duality relations of optimization problems. For that purpose we consider a convex mathematical program written in the form:

$$\inf\ f_0(\eta,x) \tag{II.1.1}$$

subject to :

$$g(x) \leq \xi' \tag{II.1.2}$$

$$h(x) = \xi'' \tag{II.1.3}$$

$$x \geq 0,$$

with f_0 being a real-valued saddle function on $\mathbf{R}^K \times \mathbf{R}^n$; $g(x)=(g_1(x),...,g_{L'}(x))$ with $g_{l'}(x)$ $(l'=1,...,L')$ being convex, lower semicontinuous on $\mathbf{R}^n$, and $h(x)=(h_1(x),...,h_{L''}(x))$ with $h_{l''}(x)$ $(l''=1,...,L'')$ linear in x on $\mathbf{R}^n$.

Let η denote a current given price vector that defines the cost $f_0(\eta,x)$ of a certain production plan x. We suppose that this price vector η is given externally, either by the market, or due to agreements, or due to estimates. The amount of capacities or resources required for producing plan x are given with $(g(x),h(x))$. Therefore with $\xi:=(\xi',\xi'')$ we are thinking of that amount of capacities or resources, which is available and might (or have to) be used. Say, we are interested in a production plan x that solves (II.1) for current (η^0,ξ^0). With paying attention to the resulting costs $f(\eta^0,.)$ we have to seek for a plan x^0, that uses resources as much as available, namely ξ^0, and that simultaneously minimizes the costs $f(\eta^0,.)$ subject to an externally given price vector η^0. With having the solution x^0 and hence the (minimal) costs $\phi(\eta^0,\xi^0)$, we are interested in the impact of (possibly external) changes $\Delta\eta$ (in the price vector η^0) on the optimal (cost) value function $\phi(.,\xi^0)$. On the other hand, having information of to what extent changes $\Delta\xi$ in the amount of available resources impact the minimal costs $\phi(\eta^0,.)$, we certainly would be able to decide up to what price limit we would buy (or sell) additional (or less efficiently used) resources.

Focusing on the optimal (cost) value function $\phi(.,.)$ at (η^0,ξ^0) we concentrate on those (sub)gradients, that yield the desired information of the socalled 'marginal costs' with respect to slight componentwise changes of (η^0,ξ^0). We denote the 'marginal costs' with (η^*,ξ^*) and may interprete them as follows :

let $\Delta\xi_l$ represent the change in the amount of available resource l; then $\xi_l^*.\Delta\xi_l$ is approximately the associated change of the (minimal) costs ϕ at (η^0, ξ^0); ξ_l^* (for all l) depend completely on internal data that characterize problem (II.1); due to this fact, ξ_l^* is often denoted as 'shadow price' for the underlying resource l ;

similarly, $\eta_k^*.\Delta\eta_k$ represents approximately that change of ϕ at (η^0,ξ^0) that is due to change $\Delta\eta_k$ of the k-th price component; again η_k^* (for all k) are determined completely by internal data of (II.1); analogously, one may think with η_k^* on a 'shadow price' that corresponds to the k-th price component.

Hence, we have to take into account that the currently given values (η^0,ξ^0) can change due to external influences (e.g. different ratings by the markets, expiring agreements, outcomes deviating from estimate) or due to internal influences (e.g. if management is thinking of restructurements). This fact strongly recommends to investigate the associated impact on the optimal cost function ϕ. Thus, we may view (η^*,ξ^*) as a price system that serves to measure the vulnerability of 'optimal' decisions with respect to an external price vector η^0 and with respect to the available capacity ξ^0, and might help to hedge against future changes in the environment. We therefore have a direct economic motivation for studying (sub)differentiability and continuity of optimal value functions corresponding to parametric optimization problems of type (II.1).

Throughout this chapter we consider the following (primal) finite dimensional convex optimization problem parametric in (η,ξ) :

$$\phi(\eta,\xi) = \inf_x \; f(\eta,\xi,x) \tag{II.2}$$

where sofar, $f(\eta,\xi,x)$ is assumed to be an extended-real-valued concave-convex saddle function being defined on $\mathbf{R}^K\times\mathbf{R}^{L+n}$ (i.e. concave in $\eta\in\mathbf{R}^K$ for all $(\xi,x)\in\mathbf{R}^{L+n}$, and convex in (ξ,x) for all η). $\phi(\eta,\xi)$ is the optimal value function. We will concentrate on (semi)continuity and subdifferentiability properties of the optimal value function $\phi(\eta,\xi)$. In doing that we strongly make use of the theory of saddle functions due to Rockafellar1964 [114]. In a certain sense our investigations relate to those of McLinden 1973 [95] discussed in the final part of this chapter. However, that which follows is new insofar as it generalizes the well-known duality results of convex optimization under convex perturbations to the case under concave-convex perturbations, not stated in literature up to now. (Below for convenience, we shall speak of the '*purely convex case*', if we are referring to convex optimization under convex perturbation.)

For the sake of exposition we restate the properties of Proposition 5.1 more generally:

Proposition II.1 : Let $f(\eta,\xi,x)$ be a concave-convex saddle-function on $\mathbf{R}^K\times\mathbf{R}^{L+n}$. Then $\phi(\eta,\xi)=\inf_x f(\eta,\xi,x)$ is a saddle function (concave-convex) on $\mathbf{R}^K\times\mathbf{R}^L$.

Proof: As the intersection of convex hypographs - characterized in taking the infimum of a collection of concave functions for fixed ξ - yields a convex hypograph, we get concavity of $\phi(.,\xi)$ in η; on the other hand keeping η fixed and minimizing a convex function under projection, yields convexity of $\phi(\eta,.)$ in ξ (see Rockafellar 1970 [118]). //

Remark II.1: Concavity of the optimal value function in the (parametric) price vector occurs already in linear programming where one takes the infimum among a collection of linear (in η) functions. Due to the first argument in the proof above, it becomes apparent that 'linearity in η' may be substituted more generally by 'concavity in η'.

Substituting $\xi=(\xi',\xi'')$, $L:=L'+L''$ in (II.1), and denoting with $X(\xi)$ the feasibility set of (II.1), with $\mathbf{I}_X$ the indicator function associated with X, we immediately see that (II.1) is of the type (II.2) with setting

$$f(\eta,\xi,x) := f_0(\eta,x) + \mathbf{I}_{X(\xi)}(x) \quad . \tag{II.3}$$

Remark II.2 : As we focus on finite dimensional optimization problems there is no need to speak of *paired topological spaces*; we simply identify the linear functions on, say, $\mathbf{R}^n$ with elements of $\mathbf{R}^n$ in terms of the inner product denoted with $\langle .,. \rangle$. Whether the presented results throughout sections 9 and 10 are extendable to infinite dimensional problems remains to be outlined.

We start (section 8) with recalling the characterization of saddle functions due to Rockafellar 1964 [114], concentrating mainly on the results that will be of importance for our investigations. In section 9 we derive a dual pair of collections of optimization problems and state assumptions under which weak and strong duality holds. In section 10 we proceed with introducing solvability, normality and stability for a dual pair of optimization problems being distinguished elements of the previously mentioned collections. We conclude that section with stating a stability criterion that ensures the existence of dual solutions. Further, we illustrate the relations of our results to those of McLinden 1973 [75] (section 11). The chapter will be concluded with applying the achieved results to convex mathematical programs (section 12).

Throughout this section we treat $\phi(\eta,\xi)$ as an arbitrary extended-real-valued saddle function being defined on $\mathbf{R}^K \times \mathbf{R}^L$, concave in η and convex in ξ. Due to Rockafellar 1964 [114] (see also Rockafellar 1970 [118, chapters 30-33]) we distinguish *proper*, *simple* and *closed* saddle functions. For the ease of exposition we define basic notions successively and in the order being needed:

$$\mathrm{dom}_\eta \, \phi := \{\eta \mid \phi(\eta,\xi) > -\infty, \forall\xi\} \ , \qquad (8.1.1)$$

$$\mathrm{dom}_\xi \, \phi := \{\xi \mid \phi(\eta \ \xi) < +\infty, \forall\eta\} \ , \qquad (8.1.2)$$

$$\mathrm{dom}_\eta' \, \phi := \{\eta \mid \phi(\eta,\xi) > -\infty, \forall\xi \in \mathrm{dom}_\xi \, \phi\} \ , \qquad (8.1.3)$$

$$\mathrm{dom}_\xi' \, \phi := \{\xi \mid \phi(\eta,\xi) < +\infty, \forall\eta \in \mathrm{dom}_\eta \, \phi\} \ . \qquad (8.1.4)$$

Obviously, $\mathrm{dom}_\eta\phi \subset \mathrm{dom}_\eta'\phi$, $\mathrm{dom}_\xi\phi \subset \mathrm{dom}_\xi'\phi$ and all are convex. Moreover, $\mathrm{dom}_\xi'\phi$, $\mathrm{dom}_\eta'\phi$ respectively, represent the effective domain of the convex function $\phi(\eta,.)$ for all $\eta \in \mathrm{ri}(\mathrm{dom}_\eta\phi)$, of the concave function $\phi(.,\xi)$ for all $\xi \in \mathrm{ri}(\mathrm{dom}_\xi \, \phi)$ respectively ([114 , Lemma 1]).

Definition 8.1 ([114],[118]) : If $(\mathrm{dom}_\eta\phi \times \mathrm{dom}_\xi'\phi) \neq \emptyset$, $(\mathrm{dom}_\eta'\phi \times \mathrm{dom}_\xi\phi) \neq \emptyset$ then ϕ is said to be **lower proper**, **upper proper** respectively. If ϕ is both, lower and upper proper, ϕ is called a **proper** saddle function. The restriction of ϕ to the sets $\mathrm{ri}(\mathrm{dom}_\eta\phi) \times \mathrm{ri}(\mathrm{dom}_\xi'\phi)$, to $\mathrm{ri}(\mathrm{dom}_\eta'\phi) \times \mathrm{ri}(\mathrm{dom}_\xi\phi)$ respectively, is denoted **lower kernel**, **upper kernel** respectively. If lower and upper kernel coincide, then ϕ is said to be **simple**, and one speaks of the **kernel** of ϕ. The function obtained by closing $\phi(\eta,\xi)$ as a concave function for each fixed ξ is called the **concave closure** of ϕ and denoted $\mathbf{cl}_\eta\phi$. Similarly, the function obtained by closing $\phi(\eta,\xi)$ as a convex function for each η is called the **convex closure** of ϕ and denoted $\mathbf{cl}_\xi\phi$. If $\phi(\eta,\xi)$ coincides with its convex closure $\mathrm{cl}_\xi\phi$, $\phi(\eta,\xi)$ is said to be **convex-closed**; analogously, if $\phi(\eta,\xi) = \mathrm{cl}_\eta\phi(\eta,\xi)$ then $\phi(\eta,\xi)$ is called **concave-closed**.

We may state in accordance with [114, Lemma 2]:

Proposition 8.1: For any saddle function ϕ it holds that $\mathrm{cl}_\eta\phi$, $\mathrm{cl}_\xi\phi$ are simple; the kernel of $\mathrm{cl}_\eta\phi$, of $\mathrm{cl}_\xi\phi$ respectively, is the upper kernel, lower kernel, respectively, of ϕ.

In dealing with saddle functions one is mostly interested in the lower saddle value (i.e. $\sup_\eta \inf_\xi \phi$), in the upper saddle value (i.e. $\inf_\xi \sup_\eta \phi$) respectively, and whether they are the same. One always has

$$\sup_\eta \inf_\xi \phi \leq \inf_\xi \sup_\eta \phi . \tag{8.2}$$

Rockafellar defined a weak partial ordering of the set of all saddle functions on $\mathbf{R}^K \times \mathbf{R}^L$ by means of the socalled *minimax extension* :

Definition 8.2 [114]: Saddle function ϕ_2 is called a **minimax extension** of saddle function ϕ_1 iff

$$\forall\ \xi, \eta^* :\ \sup_\eta \{ \phi_2(\eta,\xi) - \langle\eta,\eta^*\rangle \} \leq \sup_\eta \{ \phi_1(\eta,\xi) - \langle\eta,\eta^*\rangle \} \tag{8.3.1}$$
$$\forall\ \eta, \xi^* :\ \inf_\xi \{ \phi_2(\eta,\xi) - \langle\xi,\xi^*\rangle \} \geq \inf_\xi \{ \phi_1(\eta,\xi) - \langle\xi,\xi^*\rangle \} . \tag{8.3.2}$$

In view of (8.3.1) we conclude that the hypograph of the partial conjugate (w.r.t. η) of ϕ_2 covers the associated one of ϕ_1; analogously in (8.3.2), where the epigraph of the partial conjugate (w.r.t. ξ) of ϕ_2 covers the associated one of ϕ_1 .

Hence, if ϕ_2 is a minimax extension of ϕ_1 this entails

$$\sup_\eta \inf_\xi \phi_1 \leq \sup_\eta \inf_\xi \phi_2 \leq \inf_\xi \sup_\eta \phi_2 \leq \inf_\xi \sup_\eta \phi_1 . \tag{8.4}$$

$cl_\eta\, \phi$ and $cl_\xi\, \phi$ are examples for minimax extensions of ϕ ([114, Lemma 2]).

Definition 8.3 [114]: ϕ_1 and ϕ_2 are said to be **minimax equivalent** in the case that both are minimax extensions of each other.

Due to the above, two saddle functions are minimax equivalent iff their partial conjugates coincide. Further, for a pair ϕ_1, ϕ_2 of minimax equivalent saddle functions, it holds that for any η^*, ξ^*, the saddle functions

$$\phi_1(\eta,\xi) - \langle\eta,\eta^*\rangle - \langle\xi,\xi^*\rangle ,\ \phi_2(\eta,\xi) - \langle\eta,\eta^*\rangle - \langle\xi,\xi^*\rangle \tag{8.5}$$

have the same upper and lower saddle values and the same saddle points (if they exist).

Definition 8.4: A saddle function will be called **closed** if it is minimax equivalent to all its minimax extensions.

We subsequently restate properties of closed saddle functions being of importance for our context :

Proposition 8.2 (Rockafellar 1964 [114 , Theorem 1a,b]):
a) Each minimax equivalence class E of closed saddle functions is an 'interval' in the following sense: There exist unique saddle functions ϕ_l , ϕ_u in E, such that ϕ belongs to E if and only if $\phi_l \leq \phi \leq \phi_u$; ϕ_l is lower closed and ϕ_u upper closed. Moreover, $cl_\eta\phi = \phi_u$ and $cl_\xi\phi = \phi_l$ for all $\phi \in E$.
b) In order that a given pair of saddle functions ϕ_l, ϕ_u be the unique lower and upper members, respectively, of some minimax equivalence class of closed saddle functions, it is necessary and sufficient that $cl_\eta \phi_l = \phi_u$ and $cl_\xi \phi_u = \phi_l$.

The above relations set up a one-to-one correspondence between the (minimax) equivalence classes of closed saddle functions on $\mathbf{R}^K \times \mathbf{R}^L$ and the closed convex functions on $\mathbf{R}^{K+L}$ (Rockafellar 1967 [115]).
According to the property of minimax equivalent saddle functions, we have that all the elements of a (minimax) equivalence class of closed saddle functions have the same upper and lower saddle values and saddle points (if they exist).

Proposition 8.3 (Rockafellar 1964 [114, Corollary 2]: A saddle function ϕ is closed if and only if ϕ is minimax equivalent to $cl_\eta\phi$ and $cl_\xi\phi$.

Proposition 8.4 ([114, Corollary 3]): A saddle function ϕ is closed and proper if and only if it satisfies the following conditions:

i) $dom_\eta\phi \neq \emptyset$ and $dom_\xi\phi \neq \emptyset$; (8.6.1)
ii) $\phi(\eta,\xi) = +\infty$ when $\eta \in dom_\eta\phi$ but $\xi \notin cl(dom_\xi \phi)$; (8.6.2)
iii) $\phi(\eta,\xi) = -\infty$ when $\xi \in dom_\xi \phi$ but $\eta \notin cl(dom_\eta \phi)$; (8.6.3)
iv) $\forall \eta \in ri(dom_\eta \phi)$, $\phi(\eta,.)$ is a proper closed convex function; (8.6.4)
v) $\forall \xi \in ri(dom_\xi \phi)$, $\phi(.,\xi)$ is a proper closed concave function . (8.6.5)

Definition 8.5 (Rockafellar 1970 [118, p.359]): $cl_\xi cl_\eta\phi(\eta,\xi)$ is called the **lower closure** of $\phi(\eta,\xi)$ and $cl_\eta cl_\xi\phi(\eta,\xi)$ the **upper closure** of $\phi(\eta,\xi)$. $\phi(\eta,\xi)$ is said to be **lower closed** iff $\phi(\eta,\xi)=cl_\xi cl_\eta\phi(\eta,\xi)$; analogously, $\phi(\eta,\xi)$ is said to be **upper closed** iff $\phi(\eta,\xi)=cl_\eta cl_\xi\phi(\eta,\xi)$.

Lower and upper closures of non-closed saddle functions help to characterize the induced minimax equivalence class according to :

Proposition 8.5 (Rockafellar 1964 [114, Theorem 2]): Every saddle function ϕ has closed minimax extensions. In order that all the closed minimax extensions of ϕ be minimax equivalent, it is both necessary and sufficient that ϕ be simple. If ϕ is simple the lower and upper members of its equivalence class of closed minimax extensions are $\phi_l = cl_\xi cl_\eta \phi$, $\phi_u = cl_\eta cl_\xi \phi$.

In particular, any $\phi(\eta,\xi)$ with $\text{int}(\text{dom}_\eta \phi \times \text{dom}_\xi \phi) \neq \emptyset$ is simple ([114, p.157]); further, ϕ closed implies ϕ simple. To illustrate proper, simple and closed, we consider the following examples:

Example 8.1 (Rockafellar 1970 [118, p.361]): For $\phi: \mathbf{R} \times \mathbf{R} \to \mathbf{R}$ being defined by

$$\phi(\eta,\xi) = \begin{cases} +\infty & \text{if} \quad \eta.\xi > 0 \\ 0 & \text{if} \quad \eta.\xi = 0 \\ -\infty & \text{if} \quad \eta.\xi < 0 \end{cases}$$

we have $\text{dom}_\eta \phi = \text{dom}_\xi \phi = \{0\}$, $\text{dom}_\eta ' \phi = \text{dom}_\xi ' \phi = \mathbf{R}$ and so ϕ proper but not simple.

Example 8.2 (Rockafellar1964 [114, p.166]): For $\phi: \mathbf{R} \times \mathbf{R} \to \mathbf{R}$ being defined by

$$\phi(\eta,\xi) = \begin{cases} \eta.\xi & \text{if} \quad \eta>0,\ \xi \geq 0 \\ +\infty & \text{if} \quad \eta>0,\ \xi < 0 \\ -\infty & \text{if} \quad \eta \leq 0 \end{cases}$$

we have $\text{dom}_\eta \phi = \text{dom}_\eta ' \phi = (0,\infty)$, $\text{dom}_\xi \phi = \text{dom}_\xi ' \phi = [0,\infty)$ and hence simple; with $cl_\xi cl_\eta \phi \neq cl_\xi \phi$, ϕ is obviously not closed.

Example 8.3 (Rockafellar 1970 [118, p.360]): For $\phi: \mathbf{R} \times \mathbf{R} \to \mathbf{R}$ being given by

$$\phi(\eta,\xi) = \begin{cases} \eta^\xi & \text{if } (\eta,\xi) \in \{[0,1] \times [0,1]\} - \{(0,0)\} \\ c & \text{arbitrary with } c \in [0,1] \text{ for } (\eta,\xi) = (0,0) \end{cases}$$

and extended correspondingly with values $+\infty$, $-\infty$ outside the unit box, one easily verifies closedness via $cl_\eta cl_\xi \phi = cl_\eta \phi$, $cl_\xi cl_\eta \phi = cl_\xi \phi$.

Conjugates of a saddle function are defined according to [114]

$$\phi_l^*(\eta^*,\xi^*) := \sup_\eta \inf_\xi \{ \langle \eta,\eta^* \rangle + \langle \xi,\xi^* \rangle - \phi(\eta,\xi) \} \tag{8.7.1}$$

$$\phi_u^*(\eta^*,\xi^*) := \inf_\xi \sup_\eta \{ \langle \eta,\eta^* \rangle + \langle \xi,\xi^* \rangle - \phi(\eta,\xi) \} . \tag{8.7.2}$$

Definition 8.6 : ϕ_l^* and ϕ_u^* given through (8.7.1), (8.7.2) are called **lower, upper conjugates** of ϕ. Any saddle function ϕ^* with $\phi_l^* \le \phi^* \le \phi_u^*$ is a **conjugate** of ϕ.

The relations between an equivalence class of closed saddle functions and their conjugates are restated according to [114, Theorem 4, p.164]:

Proposition 8.6: If ϕ is a closed saddle function, then the lower and upper conjugates ϕ_l^* and ϕ_u^* of ϕ are again closed saddle functions and depend only on the minimax equivalence class containing ϕ. In fact ϕ_l^* and ϕ_u^* are the lower and upper members respectively, of a minimax equivalence class of closed saddle functions. If ϕ^* is any member of this equivalence class, the lower and upper conjugates ϕ_l^{**} and ϕ_u^{**} are in turn the lower and upper members of the closed minimax equivalence class containing ϕ; i.e. $\phi_l^{**}=\phi_l$, $\phi_u^{**}=\phi_u$. If ϕ is simple, then the closed minimax extensions of ϕ_l^* and ϕ_u^* are all minimax equivalent, and their conjugates in turn give the class of closed minimax extensions of ϕ.

Let $\partial_\eta\phi(\eta_0,\xi_0)$ denote the set of subgradients at η_0 of the concave function $\phi(.,\xi_0)$ and $\partial\phi_\xi(\eta_0,\xi_0)$ the set of subgradients at ξ_0 of the convex function $\phi(\eta_0,.)$.

Definition 8.7: $\partial\phi(\eta_0,\xi_0)=\partial_\eta\phi(\eta_0,\xi_0)\times\partial\phi_\xi(\eta_0,\xi_0)$ is called the **subdifferential at** (η_0,ξ_0), i.e. the set of subgradients of ϕ at (η_0,ξ_0).

Recalling Lemma 5, Theorem 6 in Rockafellar 1964 [114, p.169,170] we conclude this section with stating two propositions on subdifferentiability:

Proposition 8.7 : Let ϕ be a proper closed saddle function. Then :

$$\partial_\eta\phi(\eta_0,\xi_0) \neq \emptyset \text{ for all } \xi_0 \text{ whenever } \eta_0 \in \mathrm{ri}(\mathrm{dom}_\eta\,\phi)\,, \tag{8.8.1}$$
$$\partial_\xi\phi(\eta_0,\xi_0) \neq \emptyset \text{ for all } \eta_0 \text{ whenever } \xi_0 \in \mathrm{ri}(\mathrm{dom}_\xi\,\phi)\,; \tag{8.8.2}$$
$$\partial\phi(\eta_0,\xi_0) \neq \emptyset \text{ if } (\eta_0,\xi_0)\in \mathrm{ri}(\mathrm{dom}_\eta\phi)\times\mathrm{ri}(\mathrm{dom}_\xi\phi)\,. \tag{8.8.3}$$

Proposition 8.8: Let ϕ and ϕ^* be closed saddle functions conjugate to one another. Then the following conditions are equivalent:

i) $(\eta_0^*,\xi_0^*) \in \partial\phi(\eta_0,\xi_0)$; (8.9.1)

ii) $(\eta_0,\xi_0) \in \partial\phi^*(\eta_0^*,\xi_0^*)$; (8.9.2)

iii) (η_0,ξ_0) is a saddle point of $\phi(\eta,\xi) - \langle\eta_0^*,\eta\rangle - \langle\xi_0^*,\xi\rangle$; (8.9.3)

iv) (η_0^*,ξ_0^*) is a saddle point of $\phi^*(\eta^*,\xi^*) - \langle\eta^*,\eta_0\rangle - \langle\xi^*,\xi_0\rangle$. (8.9.4)

Having in mind problem (II.2) and the results of the previous section, we consider

$$\phi_f(\eta,\xi) = \inf_x f(\eta,\xi,x) \qquad (9.1)$$

where $f:\mathbf{R}^K \times \mathbf{R}^{L+n} \to \mathbf{R} \cup \{-\infty,+\infty\}$ is supposed to be an extended-real-valued, closed, concave-convex saddle function. Therefore, with having an equivalence class $E:=[cl_\xi f, cl_\eta f]$ of closed saddle functions we stress the dependence of the optimal value function on f in E with writing ϕ_f. In the following we shorten $[cl_\xi f, cl_\eta f]$ to $[f_l, f_u]$, where f_l denotes the lower closed and f_u the upper closed element of E. Unfortunately closedness of $f(\eta,\xi,x)$ - similarly to the purely convex case - do not imply closedness of ϕ .

Remark 9.1: In the case that $f(\eta,\xi,x)$ is only simple, we have to perform a regularization with passing over to its closed minimax extensions, given by $[cl_\xi cl_\eta f,\ cl_\eta cl_\xi f]$, which are minimax equivalent due to Proposition 8.5. If $f(\eta,\xi,x)$ is even not simple, we have to consider $cl_\xi f$ or $cl_\eta f$ both being simple (Proposition 8.1) and to concentrate on one of the induced equivalence classes - in particular either on $[cl_\xi cl_\eta f\ ,\ cl_\eta cl_\xi cl_\eta f]$ or $[cl_\xi cl_\eta cl_\xi f\ ,\ cl_\eta cl_\xi f]$ - of closed saddle functions.

Due to the generic nature of Fenchel-Rockafellar's duality theory, deriving the dual problem associated to a (primal) parameterized optimization problem is done via the Lagrangians (see e.g. Rockafellar 1974 [119]). We keep on going this trace and define **upper** and **lower Lagrangians** of $f(\eta,\xi,x)$ according to:

$$L_{u,f}(\eta^*,\xi^*,x) := \inf_\xi \sup_\eta [\ \langle\eta^*,\eta\rangle + \langle\xi^*,\xi\rangle + f(\eta,\xi,x)] \qquad (9.2.1)$$

$$L_{l,f}(\eta^*,\xi^*,x) := \sup_\eta \inf_\xi [\ \langle\eta^*,\eta\rangle + \langle\xi^*,\xi\rangle + f(\eta,\xi,x)] \qquad (9.2.2)$$

with obviously $L_{u,f}(\eta^*,\xi^*,x)$ being closed in ξ^* and $L_{l,f}(\eta^*,\xi^*,x)$ being closed in η^*; both are convex in (η^*,x) and concave in ξ^*. The Lagrangian might be interpreted as that function which states the value for stationarity with respect to presumed (internal) marginal costs (η^*,ξ^*) and with respect to a given decision x (more precisely with respect to $f(.,.,x)$) in the price vector η and in capacity ξ. The Lagrangians in (9.2.1) and (9.2.2) might differ due to the priority posed on the stationarity in η versus the stationarity in ξ.

Remark 9.2: 'f' as subscript of the Lagrangian indicates the dependence on the member of $[f_l,f_u]$; 'l', 'u' as subscripts of 'L' indicate the 'sup inf', 'inf sup' operation in (9.2.1) and (9.2.2).

Next, minimizing $L_{u,f}(\eta^*,\xi^*,x)-<x^*,x>$, $L_{l,f}(\eta^*,\xi^*,x)-<x^*,x>$ in x we get

$$g_u(\eta^*,\xi^*,x^*):= \inf_x [L_{u,f}(\eta^*,\xi^*,x)-<x^*,x>] \quad (9.3.1)$$
$$g_f(\eta^*,\xi^*,x^*) := \inf_x [L_{l,f}(\eta^*,\xi^*,x)-<x^*,x>] \quad (9.3.2)$$

both being closed: g_u is the upper conjugate of [-f], more precisely $g_u(\eta^*,\xi^*,x^*):= [-f]_u^*(\eta^*,\xi^*,-x^*)$, is closed due to Proposition 8.6 with $f(\eta,\xi,x)$ supposed to be closed; g_f is closed as with g_l being defined according to

$$g_l\,(\eta^*,\xi^*,x^*):= [-f]_l^*(\eta^*,\xi^*,-x^*) \quad (9.4)$$

(i.e. the lower conjugate of [-f]), we have $g_l \le g_f \le g_u$ and hence g_f is element of the equivalence class of closed saddle functions being conjugate to [-f].

Remark 9.3: 'f' as subscript of 'g' indicates the dependence on the member of $[f_l,f_u]$; 'l', 'u' as subscripts of 'g' indicate the unique lower, upper member of the equivalence class of [-f]*.

Interpreting any member $g(\eta^*,\xi^*,x^*)$ of its equivalence class as a parametric (in x*) convex-concave saddle function we turn our interests to the associated upper and lower saddle value function for $(\eta^*,\xi^*)\to g(\eta^*,\xi^*,x^*)$. As with fixing x* closedness of $(\eta^*,\xi^*)\to g(\eta^*,\xi^*,x^*)$ might be lost, we concentrate separately on the $g_u(\eta^*,\xi^*,x^*)$, $g(\eta^*,\xi^*,x^*)$, $g_l(\eta^*,\xi^*,x^*)$ being given through (9.3.1),(9.4) and define upper and lower saddle value functions $\gamma_{u,u}$, $\gamma_{u,l}$, $\gamma_{g,u}$, $\gamma_{g,l}$, $\gamma_{l,u}$, $\gamma_{l,l}$:

$$\gamma_{u,u}(x^*) := \inf_{\eta^*} \sup_{\xi^*} g_u(\eta^*,\xi^*,x^*) \quad (9.5.1)$$
$$\gamma_{u,l}(x^*) := \sup_{\xi^*} \inf_{\eta^*} g_u(\eta^*,\xi^*,x^*) \quad (9.5.2)$$
$$\gamma_{g,u}(x^*) := \inf_{\eta^*} \sup_{\xi^*} g(\eta^*,\xi^*,x^*) \quad (9.5.3)$$
$$\gamma_{g,l}(x^*) := \sup_{\xi^*} \inf_{\eta^*} g(\eta^*,\xi^*,x^*) \quad (9.5.4)$$
$$\gamma_{l,u}(x^*) := \inf_{\eta^*} \sup_{\xi^*} g_l(\eta^*,\xi^*,x^*) \quad (9.5.5)$$
$$\gamma_{l,l}(x^*) := \sup_{\xi^*} \inf_{\eta^*} g_l(\eta^*,\xi^*,x^*)\ . \quad (9.5.6)$$

We observe:

Lemma 9.1: $\forall g\in[g_l,g_u]$: $\gamma_{u,u}(x^*),\gamma_{u,l}(x^*),\gamma_{g,u}(x^*),\gamma_{g,l}(x^*),\gamma_{l,u}(x^*),\gamma_{l,l}(x^*)$ are concave in x^*. Further, $\gamma_{u,l}(x^*)\leq\gamma_{u,u}(x^*)$, $\gamma_{l,l}(x^*)\leq\gamma_{l,u}(x^*)$, $\gamma_{g,u}(x^*)\in[\gamma_{l,u}(x^*),\gamma_{u,u}(x^*)]$ and $\gamma_{g,l}(x^*)\in[\gamma_{l,l}(x^*),\gamma_{u,l}(x^*)]$.

Proof : Concavity follows by an analogous argumentation as in the proof of Proposition II.1; the inequalities are obvious and the last two assertions are implied by $g\in[g_l\,,\,g_u]$ which completes the proof. //

In view of the fact that for any member $f\in[f_l\,,\,f_u]$

$$\mathrm{cl}_\xi\,\mathrm{cl}_\eta\,f(0,0,x) = \sup_{\xi^*}\,\inf_{\eta^*}\,L_{u,f}(\eta^*,\xi^*,x)\quad, \tag{9.6.1}$$
$$\mathrm{cl}_\eta\,\mathrm{cl}_\xi\,f(0,0,x) = \inf_{\eta^*}\,\sup_{\xi^*}\,L_{l,f}(\eta^*,\xi^*,x)\quad, \tag{9.6.2}$$

and for g_u and some $g_f\in[g_l\,,\,g_u]$

$$g_u(\eta^*,\xi^*,0) = \inf_x\,L_{u,f}(\eta^*,\xi^*,x)\quad, \tag{9.7.1}$$
$$g_f(\eta^*,\xi^*,0) = \inf_x\,L_{l,f}(\eta^*,\xi^*,x) \tag{9.7.2}$$

we denote

$$\mathbf{inf_x\ \ f(0,0,x)} \qquad\qquad \mathbf{with}\ f\in[f_l\,,\,f_u] \tag{9.8}$$

as **primal collection** of optimization problems and with

$$\left.\begin{matrix}\mathbf{inf}_{\eta^*}\,\mathbf{sup}_{\xi^*}\\ \mathbf{sup}_{\xi^*}\,\mathbf{inf}_{\eta^*}\end{matrix}\right\}\ \mathbf{g(\eta^*,\xi^*,0)} \qquad\qquad \mathbf{with}\ \ g\in[g_l\,,\,g_u] \tag{9.9}$$

the associated **dual collection** of optimization problems (more precisely: the associated *dual collection of minmax problems*).

The above introduced dual pair of collections of optimization problems satisfy the following weak duality relations:

Theorem 9.1 (*Weak Duality*): Under the assumption of $f:\mathbf{R}^K \times \mathbf{R}^{L+n} \to \mathbf{R}$ being a closed saddle function with f_l, f_u as lower and upper elements of the induced equivalence class and g_u, g_l supposed to be defined according to (9.3.1),(9.4) then:

i) $\forall f \in [f_l, f_u]$: $\sup_{\xi^*} \inf_{\eta^*} g_l(\eta^*,\xi^*,0) \leq cl_\xi cl_\eta \, \phi_f(0,0) \leq$
$$\leq \sup_{\xi^*} \inf_{\eta^*} g_u(\eta^*,\xi^*,0) \ ; \qquad (9.10.1)$$

ii) $\forall g \in [g_l, g_u]$: $\sup_{\xi^*} \inf_{\eta^*} g(\eta^*,\xi^*,0) = \gamma_{g,l}(0) \leq$
$$\leq cl \, \gamma_{g,l}(0) = \inf_x [\, cl_{\xi,x} cl_\eta \, f(0,0,x)] \ ; \qquad (9.10.2)$$

iii) $\forall g \in [g_l, g_u]$: $\inf_{\eta^*} \sup_{\xi^*} g(\eta^*,\xi^*,0) = \gamma_{g,u}(0) \leq$
$$\leq cl \, \gamma_{g,u}(0\,) \leq \inf_x [\, cl_\eta cl_{\xi,x} \, f(0,0,x)] \ ; \qquad (9.10.3)$$

iv) $\forall f \in [f_l, f_u]$: $cl_\eta cl_\xi \, \phi_f(0,0) = \inf_{\eta^*} \sup_{\xi^*} g_l(\eta^*,\xi^*,0) = \gamma_{l,u}(0)$. (9.10.4)

Proof:

i) Due to closedness of g, we may write ($\forall g^* \in [g_l^*, g_u^*]$) :

$$\begin{aligned}
&g_l(\eta^*,\xi^*,0) = cl_{\eta^*} cl_{\xi^*} cl_{x^*} g(\eta^*,\xi^*,0) = [g]_l^{**}(\eta^*,\xi^*,0) = \\
&\sup_\eta \inf_\xi \inf_x [\, \langle \eta^*,\eta \rangle + \langle \xi^*,\xi \rangle + \langle 0,x \rangle - g^*(\eta,\xi,x)] = \\
&\sup_\eta \inf_\xi \inf_x [\, \langle \eta^*,\eta \rangle + \langle \xi^*,\xi \rangle + \langle 0,-x \rangle - g^*(\eta,\xi,-x)] = \qquad (9.11) \\
&\sup_\eta \inf_\xi \inf_x [\, \langle \eta^*,\eta \rangle + \langle \xi^*,\xi \rangle + \langle 0,-x \rangle + [-g]^*(-\eta,-\xi,x)] = \\
&\sup_\eta \inf_\xi \inf_x [\, \langle \eta^*,\eta \rangle + \langle \xi^*,\xi \rangle + \langle 0,-x \rangle + f(\eta,\xi,x)] = \\
&\sup_\eta \inf_\xi [\, \langle \eta^*,\eta \rangle + \langle \xi^*,\xi \rangle + \inf_x f(\eta,\xi,x)] \quad \forall f \in [f_l\,,\, f_u] \ .
\end{aligned}$$

Hence we get $\forall f \in [f_l\,,\, f_u]$:

$$\begin{aligned}
&\sup_{\xi^*} \inf_{\eta^*} g_l(\eta^*,\xi^*,0) = \\
&= \sup_{\xi^*} \inf_{\eta^*} [\sup_\eta \inf_\xi [\langle \eta^*,\eta \rangle + \langle \xi^*,\xi \rangle + \inf_x f(\eta,\xi,x)]] \leq \qquad (9.12) \\
&\leq \sup_{\xi^*} \inf_{\eta^*} \inf_\xi \sup_\eta [\langle \eta^*,\eta \rangle + \langle \xi^*,\xi \rangle + \inf_x f(\eta,\xi,x)] = \\
&= \sup_{\xi^*} \inf_\xi \inf_{\eta^*} \sup_\eta [\langle \eta^*,\eta \rangle + \langle \xi^*,\xi \rangle + \inf_x f(\eta,\xi,x)] = \\
&= \sup_{\xi^*} \inf_\xi \inf_{\eta^*} \sup_\eta [\langle \eta^*,\eta-0 \rangle + \langle \xi^*,\xi-0 \rangle + \phi_f(\eta,\xi)] = cl_\xi cl_\eta \, \phi_f(0,0) \ ,
\end{aligned}$$

proving the first inequality; similarly for the second we argue $\forall f \in [f_l\,,\, f_u]$:

$$\begin{aligned}
&\sup_{\xi^*} \inf_{\eta^*} g_u(\eta^*,\xi^*,0) = \sup_{\xi^*} \inf_{\eta^*} \inf_x L_{u,f}(\eta^*,\xi^*,x)] = \\
&= \sup_{\xi^*} \inf_{\eta^*} \inf_x \inf_\xi \sup_\eta [<\eta^*,\eta> + <\xi^*,\xi> + f(\eta,\xi,x)] \geq \\
&\geq \sup_{\xi^*} \inf_{\eta^*} \inf_\xi \sup_\eta [<\eta^*,\eta> + <\xi^*,\xi> + \inf_x f(\eta,\xi,x)] = \qquad (9.13)\\
&= \sup_{\xi^*} \inf_\xi \inf_{\eta^*} \sup_\eta [<\eta^*,\eta-0> + <\xi^*,\xi-0> + \inf_x f(\eta,\xi,x)] = \\
&= \sup_{\xi^*} \inf_\xi \inf_{\eta^*} \sup_\eta [<\eta^*,\eta-0> + <\xi^*,\xi-0> + \phi_f(\eta,\xi)] = \\
&= cl_\xi cl_\eta\ \phi_f(0,0) \ \ \forall f \in [f_l , f_u] .
\end{aligned}$$

ii) The first relation in (9.10.2) is definition (9.5.4) for x*=0 and the second one due to the concavity of the lower saddle value functions (Lemma 9.1); we concentrate on proving the last relation: Due to closedness of $f(\eta,\xi,x)$ and due to Proposition 8.6 we may conclude :

$$\begin{aligned}
&cl_{\xi,x}\ cl_\eta\ f(0,0,x) = f_l^{**}(0,0,x) = \\
&= \sup_{x^*} \sup_{\xi^*} \inf_{\eta^*} [<x,x^*> + <0,\eta^*> + <0,\xi^*> - f^*(\eta^*,\xi^*,x^*)] \\
&= \sup_{x^*} \sup_{\xi^*} \inf_{\eta^*} [<x,x^*> + <0,-\eta^*> + <0,-\xi^*> - f^*(-\eta^*,-\xi^*,x^*)] \quad (9.14)\\
&= \sup_{x^*} \sup_{\xi^*} \inf_{\eta^*} [<x,x^*> + <0,-\eta^*> + <0,-\xi^*> +[-f]^*(\eta^*,\xi^*,-x^*)] \\
&= \sup_{x^*} [<x,x^*> + \sup_{\xi^*} \inf_{\eta^*} g(\eta^*,\xi^*,x^*)] = [-\gamma_{g,l}]^*(x) \quad ,
\end{aligned}$$

with choosing g as an arbitrary member of the equivalence class corresponding to the conjugates of [-f]. We so obtain

$$cl_{\xi,x}\ cl_\eta\ f(0,0,x) = [-\gamma_{g,l}]^*(x) \qquad \forall g \in [g_l , g_u] \ . \qquad (9.15)$$

Taking next the conjugates on both sides this yields

$$\sup_x [<x,x^*> - cl_{\xi,x}\ cl_\eta\ f(0,0,x)] = [-\gamma_{g,l}]^{**}(x^*) = cl_{x^*} [-\gamma_{g,l}](x^*), \ (9.16)$$

and in particular

$$\inf_x [cl_{\xi,x}\ cl_\eta\ f(0,0,x)] = cl_{x^*} [\gamma_{g,l}](0) \qquad (9.17)$$

for any $g(\eta^*,\xi^*,x^*)$, completing the proof for ii) .

iii) The first relation is definition (9.5.3) for x*=0, the second due to the concavity of the upper saddle function, and for the third we argue as follows :

$$\begin{aligned}
&cl_\eta cl_{\xi,x}\ f(0,0,x) = f_u^{**}(0,0,x) = \\
&= \inf_{\eta^*} \sup_{x^*} \sup_{\xi^*} [<x,x^*> + <0,\eta^*> + <0,\xi^*> - f^*(\eta^*,\xi^*,x^*)] = \\
&= \inf_{\eta^*} \sup_{x^*} \sup_{\xi^*} [<x,x^*> + <0,-\eta^*> + <0,-\xi^*> - f^*(-\eta^*,-\xi^*,x^*)] = \quad (9.18)
\end{aligned}$$

$$= \inf_{\eta^*} \sup_{x^*} \sup_{\xi^*} [<x,x^*> + <0,-\eta^*> + <0,-\xi^*> +[-f]^*(\eta^*,\xi^*,-x^*)] \geq$$
$$\geq \sup_{x^*} \inf_{\eta^*} \sup_{\xi^*} [<x,x^*> + <0,-\eta^*> + <0,-\xi^*> +[-f]^*(\eta^*,\xi^*,-x^*)] =$$
$$= \sup_{x^*} [<x,x^*> + \inf_{\eta^*} \sup_{\xi^*} g(\eta^*,\xi^*,x^*)] = [-\gamma_{g,u}]^*(x) \quad \forall g \in [g_l , g_u]$$

yielding

$$cl_\eta cl_{\xi,x} f(0,0,x) \geq [-\gamma_{g,u}]^*(x) \qquad \forall g \in [g_l , g_u] . \tag{9.19}$$

Taking next the conjugates of both functions in (9.19) we get $\forall g \in [g_l , g_u]$:

$$\sup_x [<x,x^*> - cl_\eta cl_{\xi,x} f(0,0,x)] \leq [-\gamma_{g,u}]^{**}(x^*) = cl [-\gamma_{g,u}](x^*) , \tag{9.20}$$

yielding in particular

$$\inf_x [cl_\eta cl_{\xi,x} f(0,0,x)] \geq cl [\gamma_{g,u}](0) , \tag{9.21}$$

which completes the proof of iii) .

iv) From i) we have $g_l(\eta^*,\xi^*,0) = \sup_\eta \inf_\xi [<\eta^*,\eta>+<\xi^*,\xi>+\inf_x f(\eta,\xi,x)]$ for any $f \in [f_l , f_u]$. Taking the 'infsup' operation yields

$$\inf_{\eta^*} \sup_{\xi^*} g_l(\eta^*,\xi^*,0) =$$
$$= \inf_{\eta^*} \sup_{\xi^*} \sup_\eta \inf_\xi [<\eta^*,\eta> + <\xi^*,\xi> + \inf_x f(\eta,\xi,x)] = \tag{9.22}$$
$$= \inf_{\eta^*} \sup_\eta \sup_{\xi^*} \inf_\xi [<\eta^*,\eta-0> + <\xi^*,\xi-0> + \inf_x f(\eta,\xi,x)] =$$
$$= \inf_{\eta^*} \sup_\eta \sup_{\xi^*} \inf_\xi [<\eta^*,\eta-0> + <\xi^*,\xi-0> + \phi_f(\eta,\xi)] =$$
$$= cl_\eta \, cl_\xi \, \phi_f(0,0) \qquad \forall f \in [f_l , f_u] ,$$

completing the proof of iv) and hence of the Theorem. //

Under the assumptions of Theorem 9.1 we below state conclusions being of importance for introducing 'stable' and 'normal' optimization problems for a distinguished pair of (9.8) and (9.9).

Corollary 9.1.1: $g_l(\eta^*,\xi^*,0)$ is identical with the lower conjugate of $-\phi_f(\eta,\xi)$, $\forall f \in [f_l, f_u]$; i.e. $g_l(\eta^*,\xi^*,0) = [-\phi_f]_l^*$.

Proof : Follows immediately from (9.11) in the proof of part i) of Theorem 9.1. //

Corollary 9.1.2: Let $\phi_f(\eta,\xi)$ be closed for some $f\in[f_l,f_u]$. Then: $[g_l(\eta^*,\xi^*,0),\ cl_{\xi^*}\, g_l(\eta^*,\xi^*,0)]$ is an equivalence class of closed saddle functions on $\mathbf{R}^K\times\mathbf{R}^L$. Further, $cl_{\xi^*}g_l(\eta^*,\xi^*,0)\leq g_u(\eta^*,\xi^*,0)$.

Proof : Using $g_l(\eta^*,\xi^*,0)=[-\phi_f]_l^*(\eta^*,\xi^*)$ we obtain through closedness and taking cl_{ξ^*}-operation the upper closed member of the associated equivalence class with $cl_{\xi^*}g_l(\eta^*,\xi^*,0)= [-\phi_f]_u^*(\eta^*,\xi^*)$ for some $f\in[f_l,f_u]$. The last assertion is obvious. //

Corollary 9.1.3 : Assuming for some $f\in[f_l,f_u]$ that $\phi_f(\eta,\xi)$ is simple on $\mathbf{R}^K\times\mathbf{R}^L$ and $g_l(\eta^*,\xi^*,0)$ is closed on $\mathbf{R}^K\times\mathbf{R}^L$ (or a fortiori $\phi_f(\eta,\xi)$ closed on $\mathbf{R}^K\times\mathbf{R}^L$). Then the existence of a saddle value of $g_l(\eta^*,\xi^*,0)$ is equivalent with $cl_\xi cl_\eta\phi_f(0,0)=cl_\eta cl_\xi\, \phi_f(0,0)$; more precisely :

$$\sup_{\xi^*} \inf_{\eta^*} g_l(\eta^*,\xi^*,0) = cl_\xi cl_\eta\, \phi_f(0,0) \ , \tag{9.23.1}$$
$$\inf_{\eta^*} \sup_{\xi^*} g_l(\eta^*,\xi^*,0) = cl_\eta cl_\xi\, \phi_f(0,0) \ . \tag{9.23.2}$$

Proof: According to Corollary 9.1.1 we have $g_l(\eta^*,\xi^*,0)=[-\phi_f]_l^*(\eta^*,\xi^*)$ being closed. Taking next the conjugates we may conclude due to Proposition 8.6 :

$$\begin{aligned}
&\inf_{\xi^*} \sup_{\eta^*} [\ \langle\eta^*,\eta\rangle + \langle\xi^*,\xi\rangle - g_l(\eta^*,\xi^*,0)] = \\
&= \inf_{\xi^*} \sup_{\eta^*} [\ \langle\eta^*,\eta\rangle + \langle\xi^*,\xi\rangle - [-\phi_f]_l^*(\eta^*,\xi^*)] = \\
&= [-\phi_f]_u^{**}(\eta,\xi) = cl_\xi cl_\eta\ [-\phi_f(\eta,\xi)] = -\ cl_\xi cl_\eta\ \phi_f(\eta,\xi)\ ,
\end{aligned}$$

and hence in particular (9.23.1). (9.23.2) is identical with (9.10.4) and only restated for completeness. //

Corollary 9.1.4 : Suppose $g_l(\eta^*,\xi^*,0)$ is closed with $cl_{\xi^*}g_l(\eta^*,\xi^*,0)=g_u(\eta^*,\xi^*,0)$, and $\phi_f(\eta,\xi)$ being simple for some $f\in[f_l,f_u]$. Then the equivalence classes $[g_l(\eta^*,\xi^*,0),\ g_u(\eta^*,\xi^*,0)]$ and $[-cl_\eta cl_\xi\phi_f(\eta,\xi),\ -cl_\xi cl_\eta\phi_f(\eta,\xi)]$ are conjugate to one another. In this case the existence of a saddle value of $g(.,.,0)$ - for arbitrary $g\in[g_l,g_u]$ - is equivalent with $cl_\xi cl_\eta\phi_f(0,0)=cl_\eta cl_\xi\ \phi_f(0,0)$; more precisely, for some $f\in[f_l,f_u]$ and for all $g\in[g_l,g_u]$ it holds :

$$\sup_{\xi^*} \inf_{\eta^*} g(\eta^*,\xi^*,0) = cl_\xi cl_\eta\ \phi_f(0,0) \tag{9.24.1}$$
$$\inf_{\eta^*} \sup_{\xi^*} g(\eta^*,\xi^*,0) = cl_\eta cl_\xi\ \phi_f(0,0) \ . \tag{9.24.2}$$

Proof : Due to the proof of Corollary 9.1.3 we have $[g_l]_u^*(\eta,\xi) = -cl_\xi cl_\eta \phi_f(\eta,\xi)$ and hence - due to closedness of $g_l(\eta^*,\xi^*,0)$ - also $[g_u]_u^*(\eta,\xi) = -cl_\xi cl_\eta \phi_f(\eta,\xi)$. Therefore, it remains to show that $[g_u]_l^*(\eta,\xi) = -cl_\eta cl_\xi \phi_f(\eta,\xi)$ or equivalently $[g_l]_l^*(\eta,\xi) = -cl_\eta cl_\xi \phi_f(\eta,\xi)$. This becomes evident through

$$\begin{aligned} &\sup_{\eta^*} \inf_{\xi^*} [<\eta^*,\eta> + <\xi^*,\xi> - g_l(\eta^*,\xi^*,0)] = \\ &= \sup_{\eta^*} \inf_{\xi^*} [<\eta^*,\eta> + <\xi^*,\xi> - [-\phi_f]_l^*(\eta^*,\xi^*)] = \\ &= [-\phi_f]_l^{**}(\eta,\xi) = cl_\eta cl_\xi [-\phi_f(\eta,\xi)] = - cl_\eta cl_\xi \phi_f(\eta,\xi) . \end{aligned}$$

From the above (9.24) follows immediately for $(\eta,\xi)=(0,0)$. //

Remark 9.4 : Note that Corollary 9.1.3 and 9.1.4 state two different types of strong duality in the case that ϕ_f is lower and upper closed at (0,0) for some f of its equivalence class. With Corollary 9.1.3 we obtain the relation

$$\sup_{\xi^*} \inf_{\eta^*} g_l(\eta^*,\xi^*,0) = \inf_{\eta^*} \sup_{\xi^*} g_l(\eta^*,\xi^*,0) = \inf_x f(0,0,x) \tag{9.25}$$

while with Corollary 9.1.4 we get

$$\sup_{\xi^*} \inf_{\eta^*} g(\eta^*,\xi^*,0) = \inf_{\eta^*} \sup_{\xi^*} g(\eta^*,\xi^*,0) = \inf_x f(0,0,x) \tag{9.26}$$

for all members g of its equivalence class.

We have illustrated in (9.1)-(9.5) how to pass from a primal collection of minimization problems to a dual collection of minmax problems. For reasons of completeness and for demonstrating that the above duality scheme is truly symmetric we perform now 'the other way around' : Starting with a closed, parametric in x*, convex-concave saddle function $g:\mathbf{R}^K \times \mathbf{R}^{L+n} \to \mathbf{R}$ we are interested in solving the parametric minmax problems

$$\gamma_{g,u}(x^*) := \inf_{\eta^*} \sup_{\xi^*} g(\eta^*,\xi^*,x^*) \tag{9.27.1}$$
$$\gamma_{g,l}(x^*) := \sup_{\xi^*} \inf_{\eta^*} g(\eta^*,\xi^*,x^*) . \tag{9.27.2}$$

We define the associated Lagrangian for any member $g(\eta^*,\xi^*,x^*)$ of its equivalence class as

$$L_g(\eta^*,\xi^*,x) := \sup_{x^*} [g(\eta^*,\xi^*,x^*) + <x,x^*>] . \tag{9.28}$$

Obviously L_g is closed in x, convex in (η^*,x) and concave in ξ^*. Taking the conjugacy of $-L_g$ with respect to (η^*,ξ^*) we get

$$f_g(\eta,\xi,x) := \sup_{\xi^*} \inf_{\eta^*} [\, L_g(\eta^*,\xi^*,x) \; -<\eta^*,\eta> - <\xi^*,\xi>] \,, \qquad (9.29.1)$$
$$f_u(\eta,\xi,x) := \inf_{\eta^*} \sup_{\xi^*} [\, L_g(\eta^*,\xi^*,x) \; -<\eta^*,\eta> - <\xi^*,\xi>] \,. \qquad (9.29.2)$$

Similarly to (9.3) both $f_g(\eta,\xi,x)$ and $f_u(\eta,\xi,x)$ are closed: $f_u(\eta,\xi,x)$ represents the upper conjugate of [-g]; more precisely we obtain closedness through $f_u(\eta,\xi,x) = [-g]^*(-\eta,-\xi,x)$; $f_g(\eta,\xi,x)$ is closed as with $f_l(\eta,\xi,x)$ being given through

$$f_l(\eta,\xi,x)=[-g]_l^*(-\eta,-\xi,x) \qquad (9.30)$$

(i.e. the lower conjugate of [-g]) we have f_g being closed (as $f_l \le f_g \le f_u$ and hence $f_g \in [f_l, f_u]$).

10 *Normal and Stable Programs*

Theorem 9.1 and the corresponding Corollaries illustrate the distinguished role of the closed (more precisely lower closed) saddle functions $f_l(\eta,\xi,x)$ and $g_l(\eta^*,\xi^*,x^*)$ within the collections of optimization problems (9.8) and (9.9).

We therefore define the **primal** optimization problem as

P(0,0) $\qquad$ $\mathbf{inf_x \; f_l(0,0,x)}$ $\qquad$ (10.1.1)

and the corresponding **dual** as

D(0) $\qquad$ $\left.\begin{matrix} \inf_{\eta^*} \sup_{\xi^*} \\ \sup_{\xi^*} \inf_{\eta^*} \end{matrix}\right\} \; g_l(\eta^*,\xi^*,0)\,.$ $\qquad$ (10.1.2)

The set of minimizers of (10.1.1) are denoted with $\Phi_l(0,0)$ and the set of saddle points of (10.1.2) with $\Gamma_l(0)$; more precisely :

$$\Phi_l(0,0) := \{x_0 \mid f_l(0,0,x_0) \le f_l(0,0,x),\ \forall x\}, \qquad (10.2.1)$$
$$\Gamma_l(0) := \{(\eta_0^*,\xi_0^*) \mid g_l(\eta^*,\xi_0^*,0) \le g_l(\eta_0^*,\xi_0^*,0) \le g_l(\eta_0^*,\xi^*,0), \forall(\eta^*,\xi^*)\}. \qquad (10.2.2)$$

Definition 10.1: P(0,0) is said to be **solvable** iff minimizers exist, i.e. $\Phi_l(0,0)\neq\varnothing$; D(0) is said to be **solvable** iff saddle points exist, i.e. $\Gamma_l(0) \neq \varnothing$.

In accordance to convex optimization problems convexly perturbed (see e.g. Ekeland and Temam 1976 [39, chapter III]) we extend the notions of *normality* and *stability* for the above stated dual pair of extremal problems. With setting $\phi_l(\eta,\xi):=\inf_x f_l(\eta,\xi,x)$ we get :

Definition 10.2: P(0,0) is **normal** iff $cl_\xi cl_\eta \phi_l(0,0)=\phi_l(0,0)=cl_\eta cl_\xi \phi_l(0,0)$ with $\phi_l(0,0)$ being finite. D(0) is **normal** iff $\gamma_{l,l}(.)$ is finite and closed at 0 with $cl\gamma_{l,l}(0)=\gamma_{l,l}(0)=\gamma_{l,u}(0)$.

Definition 10.3: P(0,0) is **stable** iff P(0,0) is normal and ϕ_l is subdifferentiable at (0,0). D(0) is **stable** iff D(0) is normal and $\gamma_{l,l}$ subdifferentiable at 0 .

Above definitions yield the following statements representing extensions of the corresponding results for purely convex optimization problems (see Ekeland and Temam 1976 [39]):

Theorem 10.1: For $\phi_l(\eta,\xi)$ being simple and $g_l(\eta^*,\xi^*,0)$ closed (or a fortiori for $\phi_l(\eta,\xi)$ being closed) it holds :

i) P(0,0) normal ⇔ D(0) normal ;
ii) P(0,0) stable ⇔ P(0,0) normal, D(0) solvable ;
iii) D(0) stable ⇔ D(0) normal, P(0,0) solvable ;
iv) P(0,0) and D(0) normal, solvable ⇔ P(0,0) , D(0) stable ⇔ P(0,0) stable and solvable ⇔ D(0) stable and solvable .

Proof:

i) Becomes evident due to (9.23.1) of Corollary 9.1.3, and due to (9.10.2) and (9.10.3) of Theorem 9.1; in particular we obtain

$$\gamma_{l,l}(0) = cl_\xi cl_\eta\, \phi_l(0,0)\ ,$$
$$\gamma_{l,l}(0) \leq cl\, \gamma_{l,l}(0) = \phi_l(0,0)\ ,$$
$$\gamma_{l,u}(0) = cl_\eta cl_\xi\, \phi_l(0,0)\ ,$$

yielding immediately the equivalence relation.

ii) In any case we have P(0,0) normal, closedness of $[cl_\xi cl_\eta\ \phi_l,\ cl_\eta cl_\xi\ \phi_l]$ due to ϕ_l being (at least) simple, closedness of $[g_l, cl_{\xi*} g_l]$. Thus it reduces to prove: ϕ_l subdifferentiable at (0,0) $\Leftrightarrow$ D(0) solvable . This is seen immediately from

$$(-\eta_0^*,-\xi_0^*)\in\partial\phi_l(0,0) \Leftrightarrow (-\eta_0^*,-\xi_0^*)\in\partial(cl_\eta cl_\xi)\phi_l(0,0) \Leftrightarrow$$
$$\Leftrightarrow (-\eta_0^*,-\xi_0^*)\in\partial(cl_\xi cl_\eta)\phi_l(0,0) \Leftrightarrow (0,0)\in\partial g_l(\eta_0^*,\xi_0^*,0)$$

where the last equivalence relation is due to Corollary 9.1.1 and Proposition 8.8. This proves the equivalence in ii) .

iii) In any case we have D(0) normal. Thus it reduces to prove: $\gamma_{l,l}$ is subdifferentiable at $x^*=0 \Leftrightarrow$ P(0,0) is solvable. This becomes evident through :

$$-x_0\in\partial\gamma_{l,l}(0) \Leftrightarrow 0\in\partial[-\gamma_{l,l}]^*(x_0) \Leftrightarrow 0\in f_l(0,0,x_0)$$

whereby the last relation is due to (9.15) in the proof of part ii) of Theorem 9.1

iv) Becomes evident through combinations of i), ii) and iii). Herewith the proof is completed. //

Remark 10.1: As it becomes evident from the proof of Theorem 10.1 we realize that similarly to the purely convex case the subdifferential of ϕ_l at (0,0) solves D(0), the subdifferential of $\gamma_{l,l}$ at 0 solves P(0,0), respectively.

Remark 10.2: Note that P(0,0) being normal, D(0) being normal respectively, imply

$$\inf_x f_l(0,0,x) = \sup_{\xi^*} \inf_{\eta^*} g_l(\eta^*,\xi^*,0) = \inf_{\eta^*} \sup_{\xi^*} g_l(\eta^*,\xi^*,0)\ . \quad (10.3)$$

Thinking on purely convex optimization problems, we know that the existence of an upper bound for the optimal value of slightly perturbed problems together with the finiteness of the optimal value of the unperturbed problem is sufficient for stability of the unperturbed problem. We broaden this idea with exploiting the saddle property :

Theorem 10.2 (*Stability Criterion*): Let $f_l(\eta,\xi,x)$ be a lower closed, concave-convex saddle function on $\mathbf{R}^K \times \mathbf{R}^{L+n}$ and $\phi_l(\eta,\xi):=\inf_x f_l(\eta,\xi,x)$. If

i) $\exists$ x' so that $f_l(.,.,x')$ is continuous and finite at (0,0), (10.4.1)

ii) $\exists$ open $U_1 \in \mathcal{N}(\eta=0)$: $\phi_l(\eta,0)$ finite $\forall\ \eta \in U_1$ (10.4.2)

hold, then P(0,0) is stable.

Proof : Due to continuity and finiteness of $f_l(.,.,x')$ at (0,0) there exists open (bounded) $U_2 \in \mathcal{N}(\eta=0)$ and $V \in \mathcal{N}(\xi=0)$ so that $\phi_l(\eta,\xi) \leq f_l(\eta,\xi,x')$ $\forall$ $(\eta,\xi) \in U_2 \times V$ with entailing boundedness from above of $\phi_l(\eta,\xi)$ on $U_2 \times V$; i.e. $\phi_l(\eta,\xi) \leq b_u$ on $U_2 \times V$ for some scalar b_u . On the other hand ii) and $U=U_1 \cap U_2$ ensure a finite lower bound b_l with $b_l \leq \phi_l(\eta,0)$ $\forall \eta \in U$. As $U \times V$ are open, we obtain $U \times V \subset \mathrm{int}[\mathrm{dom}\ \phi_l(\eta,\xi)]$ and thus $(0,0) \in \mathrm{int}[\mathrm{dom}\phi_l(\eta,\xi)]$ implying

$$\phi_l(0,0)=cl_\xi cl_\eta \phi_l(0,0)=cl_\eta cl_\xi \phi_l(0,0)$$

and further subdifferentiability of $\phi_l(\eta,\xi)$ - i.e. $\partial\phi_l(\eta,\xi) \neq \emptyset$ - due to Proposition 8.7. This yields P(0,0) being stable. //

Remark 10.3: Due to the proof we have $\phi_l(0,0)=cl_\xi cl_\eta \phi_l(0,0)=cl_\eta cl_\xi \phi_l(0,0)$ ensuring the existence of a saddle value for $g_l(\eta^*,\xi^*,0)$ (due to Corollary 9.1.3 and the observation that $[cl_\xi cl_\eta\ \phi_l,\ cl_\eta cl_\xi\ \phi_l]$ is closed).

Remark 10.4 : Theorem 10.2 may be seen as extension of its purely convex counterpart (see e.g. Propositon 2.3 in Ekeland and Temam 1976 [39, p. 53] covering the Slater condition of the (finite dimensional) case.

The above stated stability criterion yields strong duality of the form

$$\inf_x f_l(0,0,x)=\max_{\xi^*} \min_{\eta^*} g_l(\eta^*,\xi^*,0)=\min_{\eta^*} \max_{\xi^*} g_l(\eta^*,\xi^*,0). \quad (10.5)$$

In the case that primal and dual problems are solvable then we have

$$\min_x f_l(0,0,x)=\max_{\xi^*} \min_{\eta^*} g_l(\eta^*,\xi^*,0)=\min_{\eta^*} \max_{\xi^*} g_l(\eta^*,\xi^*,0). \quad (10.6)$$

In this section we introduced solvable, normal and stable optimization problems according to the existence of solutions and subgradients of the value function. These notions, their relationships (Theorem 10.1) and the stability criterion (Theorem 10.2) have been motivated by their finite-dimensional, purely convex counterparts based on the following facts[1]: starting with $\inf_x f(\xi,x)$ and defining $f_1(\eta,\xi,x):=f(\xi,x)$ for any (finite-dimensional) η we obtain a saddle function $f_1(\eta,\xi,x)$ being lower closed if and only if $cl_\xi\, cl_x\, f(\xi,x) = f(\xi,x)$, and herewith the optimization problem (10.1).

11 Relation to McLinden's Results

McLinden developed dual operations on equivalence classes of saddle functions and proved an extension of Fenchel's Duality on these classes in his works 1973 [95] and 1974 [96]. In this subsection we discuss inhowfar the duality relations of Theorem 9.1 and its Corollaries are already accomplished with the results of McLinden 1973 [95] and 1974 [96].

As it is wellknown in convex analysis (Rockafellar 1970 [118]), *addition* and *infimal convolution* of proper convex functions are dual operations with respect to the conjugacy correspondence in the following sense: taking the conjugacy of a sum of proper closed convex functions is equal to the infimal convolution of the conjugates of the convex functions involved; vice versa, taking the infimal convolution of proper closed convex functions is equal to the sum of the associated conjugates. The extension of this dual pair of operations to proper closed saddle functions under appropriate assumptions is due to McLinden 1973 [95] and denoted therein as *addition* and *extremal convolution*.

A second pair of dual operations - which will be of interest for us below - arises from combining convex functions with linear transformations in two different ways. We will concentrate on the corresponding extension to closed saddle functions due to McLinden 1973 [95], modified only in notation with respect to our context inasmuch as we treat convexity in (ξ,x), concavity in (ξ^*,x^*) respectively; additionally, for reasons of convenience let $[f]$ denote the equivalence class of closed proper saddle functions induced by $f(\eta,\xi,x)$, supposed to be closed on $\mathbf{R}^K \times \mathbf{R}^{L+n}$.

1) I owe thanks to an anonymous referee for motivating this discussion

Let us define *product linear transformations*, *forming the composition with a product linear transformation*, and *taking the image under a product linear transformation* (McLinden 1973 [95, pp 368]):

Definition 11.1: $A(u,(v_1,v_2)):=(A_1(u),A_2(v_1,v_2))=(\eta,(\xi,x))$ is called a **product linear transformation mapping** from $\mathbf{R}^{K'}\times\mathbf{R}^{L'+n'}$ into $\mathbf{R}^{K}\times\mathbf{R}^{L+n}$ iff A_1, A_2 are linear transformations and map $\mathbf{R}^{K'}$ into $\mathbf{R}^{K}$, map $\mathbf{R}^{L'+n'}$ into $\mathbf{R}^{L+n}$ respectively.

Adjoint and range of A satisfy $A^*(u^*,(v_1{}^*,v_2{}^*))=(A_1{}^*(u^*),A_2{}^*(v_1{}^*,v_2{}^*))$, $rgA=rgA_1 \times rgA_2$.

Definition 11.2 : The operation on a closed saddle function $f(\eta,\xi,x)$

$$fA : (u,(v_1,v_2)) \rightarrow f(A(u,(v_1,v_2)) \qquad (11.1)$$

is called **forming the composition of [f] with A** iff for all members of the equivalence class $f(A(u,(v_1,v_2))$ is closed and belong to a single equivalence class.

Let $v=(v_1,v_2)$ and A be a linear product transformation mapping $\mathbf{R}^{K}\times\mathbf{R}^{L+n}$ into $\mathbf{R}^{K'}\times\mathbf{R}^{L'+n'}$.

Definition 11.3 : The operation (Af) on a closed saddle function $f(\eta,\xi,x)$

$$Af_1 : (u,v) \rightarrow \sup_{\{\eta|A_1(\eta)=u\}} \inf_{\{(\xi,x)|A_2(\xi,x)=v\}} f(\eta,\xi,x) \text{ ,} \qquad (11.2.1)$$

$$Af_2 : (u,v) \rightarrow \inf_{\{(\xi,x)|A_2(\xi,x)=v\}} \sup_{\{\eta|A_1(\eta)=u\}} f(\eta,\xi,x) \text{ ,} \qquad (11.2.2)$$

is called **taking the image of [f] under A** iff the saddle functions in (11.2) are closed and belong to a single equivalence class.

The following two Propositions represent Theorems 4-6 of McLinden 1973 [95] and state that *forming the composition* and *taking the image* is a dual pair of operations for closed saddle functions.

Proposition 11.1 : Let f be a proper, closed (concave-convex) saddle function on $\mathbf{R}^{K}\times\mathbf{R}^{L+n}$. Supposing that A is a product linear transformation mapping $\mathbf{R}^{K'}\times\mathbf{R}^{L'+n'}$ into $\mathbf{R}^{K}\times\mathbf{R}^{L+n}$ with $rgA \cap ri(\text{dom } f)\neq\emptyset$. Then :

i) f(A) defines a closed proper equivalence class with $\text{dom}(f(A))=A^{-1}(\text{dom } f)$;

ii) $(f_l(A))$ and $(f_u(A))$, are lower and upper closed members of its equivalence class; the subdifferential of f(A) satisfies: $\partial(f(A))(u,(v_1,v_2))=A^*\partial f(A(u,(v_1,v_2)))$.

Denoting with $A_2(f(\eta,.,.))(v)$, the image of a convex function by a linear transformation A_2, and with $A_1(f(.,\xi,x))(u)$ the image of a concave function by a linear transformation, we restate Theorems 5,6 of McLinden 1973 [95]:

Proposition 11.2 : Let f be a proper, closed (concave-convex) saddle function on $\mathbf{R}^K \times \mathbf{R}^{L+n}$. Supposing that A is a product linear transformation mapping $\mathbf{R}^K \times \mathbf{R}^{L+n}$ into $\mathbf{R}^{K'} \times \mathbf{R}^{L'+n'}$ and $rgA^* \cap ri(dom\ f^*) \neq \emptyset$. Then: [Af] defines a proper closed (unique) equivalence class with $dom(Af) \subset A(dom\ f)$. Moreover $[Af]^*=[f^*A^*]$. The least (lower closed) and greatest (upper closed) members of [Af] are given by

$$(Af)_l(u,v) = \sup_{\{\eta | A_1(\eta)=u\}} cl(A_2(f(\eta,.,.))(v) \tag{11.3.1}$$
$$(Af)_u(u,v) = \inf_{\{(\xi,x) | A_2(\xi,x)=v\}} cl(A_1(f(.,\xi,x))(u) . \tag{11.3.2}$$

Denoting with I^K, the identity matrix of dimension $(K \times K)$, with $0^{L,n}$, $0^{n,n}$ the zero matrices of sizes $(L \times n)$, $(n \times n)$ and let $A=(A_1,A_2)$ be defined by

$$A_1 := I^K \quad , \qquad A_2 := \begin{pmatrix} I^L & 0^{L,n} \\ 0^{n,L} & 0^{n,n} \end{pmatrix} \tag{11.4}$$

we get with $u=\eta$, $v=(\xi,0)$

$$(Af)_l(\eta,\xi,0) = cl_\xi [\inf_x f(\eta,\xi,x)] \tag{11.5.1}$$
$$(Af)_u(\eta,\xi,0) = \inf_x [cl_\eta f(\eta,\xi,x)] \quad , \tag{11.5.2}$$

and in particular at $(\eta,\xi)=(0,0)$

$$(Af)_l(0,0,0)= cl_\xi [\inf_x f(0,0,x)] \tag{11.6.1}$$
$$(Af)_u(0,0,0)= \inf_x [cl_\eta f(0,0,x)] . \tag{11.6.2}$$

(11.5) illustrates that the primal collection of optimization problems in (9.8) may be characterized by taking the image of f under A being specially chosen with (11.4); additionally, due to A being symmetric (i.e. $A=A^*$) we dually get $(f^*A^*)(\eta^*,\xi^*,x^*)=f^*(\eta^*,\xi^*,0)$.

Using the notations in sections 10 and 11, $f^*(\eta^*,\xi^*,x^*)=-g(-\eta^*,-\xi^*,x^*)$, and noting that $rgA=rgA^*=\mathbf{R}^K \times \mathbf{R}^L \times \{0\}$ we Propositions 11.1, 11.2 to our context with respect to the specially chosen product linear transformation A as in (11.4) :

Proposition 11.3: Let f be a proper, closed (concave-convex) saddle function on $\mathbf{R}^K \times \mathbf{R}^{L+n}$ and $f^*(\eta^*,\xi^*,x^*)=-g(-\eta^*,-\xi^*,x^*)$. Let $\mathbf{R}^K \times \mathbf{R}^L \times \{0\} \cap \mathrm{ri}(\mathrm{dom}\text{-}g) \neq \emptyset$, then: For all members f of its equivalence class $(Af)(\eta,\xi,0)=\phi_f(\eta,\xi)$ define a unique equivalence class of proper closed saddle functions. Further, for any member g of its equivalence class, $(gA^*)(\eta^*,\xi^*,x^*)=g(\eta^*,\xi^*,0)$ define a unique equivalence class of proper closed saddle functions with $\phi_f(\eta,\xi)$ conjugate to $-g(\eta^*,\xi^*,0)$.

Remark 11.1: Note that the condition $\mathbf{R}^K \times \mathbf{R}^L \times \{0\} \cap \mathrm{ri}(\mathrm{dom}\text{-}g) \neq \emptyset$ in the above proposition may substitute the assumption in Corollary 9.1.4 due to the fact that in this case $\phi_f(\eta,\xi)$ is even closed for any member f of its equivalence class; more precisely, according to the proofs of Lemma 1 and Theorems 5,6 in McLinden 1973 [95], it is ensured that every $\phi_f(\eta,\xi)$ is within the lower and upper member of the closed equivalence class corresponding to the conjugates of $g(\eta^*,\xi^*,0)$, ensuring that apparently $\phi_f(\eta,\xi)$ is closed (Proposition 8.2). Hence McLinden's condition in general is sufficient for the equivalences to hold in Theorem 10.1 .

Considering the subdifferential formula in ii) of Proposition 11.1 we may add that with writing $\partial(f^*A^*)(\eta^*,\xi^*,x^*)=A\partial f^*(A^*(\eta^*,\xi^*,x^*))$ one implicitly assumes the subdifferentiability of f^*A^* with respect to (η^*,ξ^*,x^*); due to McLinden 1973 [95, p. 372,376] this implies a strong duality relation of the form

$$\text{maximin}\,(Af) = \text{minimax}\,(-f^*A^*) \quad . \tag{11.7}$$

This fact, applied to our context with A being given as in (11.4), yields primal and dual solvability allowing to conclude with a strong duality relation of type (10.6) - a special version of (11.7) .

According to the discussion of the above we summarize that Theorem 9.1 may be seen as broadening the strong duality results of McLinden 1973 [95] to weak duality relations among (not necessarily unique) closed saddle functions that arise with taking the image (forming the composition, respectively) of closed saddle functions subject to the specially chosen product linear transformation (11.4). We further stress that in this special case we have strengthened the weak duality relations of Theorem 9.1 successively via (9.25),(9.26), (10.5) and (10.6). Whether similar results are valid also for other product linear transformations remains to be investigated. Regarding the work of McLinden 1974 [96] we further want to mention, that the notion of stability therein refers to perturbations that characterize translations of saddle functions; within our special case (11.4) we allow perturbations - expressed with x* - that behave jointly concave with ξ^*. Whether this allows some extension of Fenchel's duality within dual minimax problems remains also to be studied.

12 Application to Convex Programming

In the final part of this chapter we focus on applying (illustratively) the above derived results to a (primal) convex program of the form

$$\inf_x f_0(\eta,x) \qquad (12.1.1)$$

$$\text{s.t. } h(x) \leq \xi \ . \qquad (12.1.2)$$

Supposing f_0 is a lower closed (concave-convex) saddle function on $\mathbf{R}^K \times \mathbf{R}^n$ whose domain has a nonempty relative interior; $h(x)=(h_1(x),...,h_L(x))$ with $h_l(x)$ $(l=1,...,L)$ is convex real-valued on $\mathbf{R}^n$; $X(\xi)$ denotes the closed feasibility set of (12.1), with $\mathbf{I}_{X(\xi)}$ the indicator function of $X(\xi)$, and define $(-\infty+\infty)= +\infty$. This yields closedness of

$$f(\eta,\xi,x) := f_0(\eta,x) + \mathbf{I}_{X(\xi)}(x) \ . \qquad (12.2)$$

Due to 'supinf' and 'infsup' operation below, let us conveniently set $(-\infty+\infty)=+\infty$, $(+\infty-\infty)=-\infty$. Upper and lower Lagrangians for (12.1) may be described in accordance with (9.2.1), (9.2.2) :

$$\begin{aligned} L_{u,f}(\eta^*,\xi^*,x) &:= \inf_\xi \sup_\eta [<\eta^*,\eta> + <\xi^*,\xi> + f(\eta,\xi,x)]= \qquad (12.3.1)\\ &= \inf_\xi \sup_\eta [<\eta^*,\eta> + <\xi^*,\xi> + f_0(\eta,x) + \mathbf{I}_{X(\xi)}(x)] = \\ &= \inf_\xi [<\xi^*,\xi> + \mathbf{I}_{X(\xi)}(x)] + \sup_\eta [<\eta^*,\eta> + f_0(\eta,x)] = \\ &= \inf_\xi [<\xi^*,\xi> + \mathbf{I}_{X(\xi)}(x)] + [-f_0]^*(\eta^*,x) \ , \end{aligned}$$

$$\begin{aligned} L_{l,f}(\eta^*,\xi^*,x) &:= \sup_\eta \inf_\xi [<\eta^*,\eta> + <\xi^*,\xi> + f(\eta,\xi,x)] \qquad (12.3.2)\\ &= \sup_\eta [<\eta^*,\eta> + f_0(\eta,x)] + [\inf_\xi [<\xi^*,\xi> + \mathbf{I}_{X(\xi)}(x)] = \\ &= [-f_0]^*(\eta^*,x) + \inf_\xi [<\xi^*,\xi> + \mathbf{I}_{X(\xi)}(x)] \ . \end{aligned}$$

Utilizing that

$$\inf_\xi \left[\langle \xi^*,\xi\rangle + I_{X(\xi)}(x)\right] = \begin{cases} \Sigma_l \ \xi_l^*.\ h_l(x) & \text{for } \xi^*\geq 0 \\ -\infty & \text{else} \end{cases} \qquad (12.4)$$

yields for $\xi^*\geq 0$,

$$L_{u,f}(\eta^*,\xi^*,x) = L_{l,f}(\eta^*,\xi^*,x) = \Sigma_l \ \xi_l^*.\ h_l(x) + [-f_0]^*(\eta^*,x) \ . \qquad (12.5)$$

In a next step we get for $\xi^*\geq 0$

$$g_u(\eta^*,\xi^*,x^*) = g_f(\eta^*,\xi^*,x^*) = \inf_x \Sigma_l \ \xi_l^*.h_l(x) + [-f_0]^*(\eta^*,x) - \langle x,x^*\rangle \qquad (12.6)$$

and in particular for $x^*=0$ and $\xi^*\geq 0$:

$$g_u(\eta^*,\xi^*,0) = g_f(\eta^*,\xi^*,0) = \inf_x \ [\Sigma_l \ \xi_l^*.h_l(x) + [-f_0]^*(\eta^*,x)]. \qquad (12.7)$$

In the case that for any member $g(\eta^*,\xi^*,x^*)$ of its equivalence class on $\mathbf{R}^K\times\mathbf{R}^{L+n}$, $g(.,.,0)$ is closed and defines a single equivalence class, then the dual minmax program is given with:

$$D(0): \quad \left.\begin{matrix} \inf_{\eta^*} \sup_{\xi^*} \\ \sup_{\xi^*} \inf_{\eta^*} \end{matrix}\right\} \ \inf_x \ [\Sigma_l \ \xi_l^*.h_l(x) + [-f_0]^*(\eta^*,x)] \qquad (12.8)$$

$$\text{s.t. } \xi^*\geq 0 \ .$$

We conclude this section with applying above relations to a parameterized convex program :

<u>Example 12.1</u> :

$$\inf_x \ f_0(\eta,x) - \langle c,x\rangle \qquad (12.9)$$

with

$$f_0(\eta,x) := \begin{cases} \Sigma_i \ \eta_i.x_i^2 + \Sigma_i \ \varepsilon_i.x_i^2 & \text{for } \eta_i \geq -\varepsilon_i \\ -\infty & \text{else} \end{cases} \qquad (12.10)$$

where ε_i (i=1,...,n) are supposed to be strictly positive. Obviously $f_0(\eta,x)$ is a proper closed (concave-convex) saddle function. As f_0 is continuous on its domain, the equivalence class (restricted to the domain) consists of one element. The upper and lower Langrangians coincide and are of the form

$$L(\eta^*,x) := \sup_\eta [<\eta^*,\eta> + f_0(\eta,x) - <c,x>] = \tag{12.11}$$
$$= \sup_{\eta\geq -\varepsilon} [<\eta^*,\eta> + \Sigma_i \eta_i.x_i^2 + \Sigma_i \varepsilon_i.x_i^2 - <c,x>] =$$
$$= \sup_{\eta\geq -\varepsilon} [\Sigma_i \eta_i.(x_i^2 + \eta_i^*) + \Sigma_i \varepsilon_i.x_i^2 - <c,x>] .$$

This yields

$$L(\eta^*,x) = \begin{cases} \Sigma_i -\varepsilon_i.\eta_i^* - <c,x> & \text{if } \forall\, i : (x_i^2 + \eta_i^*) \leq 0 \\ + \infty & \text{else} \end{cases} \tag{12.12}$$

Defining for $\eta_i^*\leq 0$, $\alpha(\eta^*)=(\alpha_1(\eta_1^*),\ldots,\alpha_n(\eta_n^*))$, $\alpha_i(\eta_i^*)=-(-\eta_i^*)^{-1/2}$ (i=1,...,n) and $\beta(\eta^*)=(\beta_1(\eta_1^*),\ldots,\beta_n(\eta_n^*))$, $\beta_i(\eta_i^*)=(-\eta_i^*)^{-1/2}$ (i=1,...,n), then the dual to (12.9) is given with

$$g(\eta^*,0) = \begin{cases} \inf_{x\in[\alpha(\eta^*),\beta(\eta^*)]} [\Sigma_i -\varepsilon_i.\eta_i^* - <c,x>] & \eta_i^* \leq 0 \\ + \infty & \text{else} \end{cases} . \tag{12.13}$$

For $\eta_i^*\leq 0$ (i=1,...,n) we obtain as optimal solution in the right-hand side of (12.13)

$$x_i^{opt}(\eta_i^*) := \begin{cases} \alpha_i(\eta_i^*) & \text{if} : c_i \leq 0 \\ \beta_i(\eta_i^*) & \text{if} : c_i > 0 \end{cases} . \tag{12.14}$$

Setting $x^{opt}(\eta^*)=(x_1^{opt}(\eta_1^*),\ldots,x_n^{opt}(\eta_n^*))$ yields for $\eta_i^* \leq 0$ (i=1,...,n)

$$g(\eta^*,0) = [\Sigma_i -\varepsilon_i.\eta_i^* - <x^{opt}(\eta^*),c>] . \tag{12.15}$$

Solving the dual problem

$$\inf_{\eta^*\leq 0} \Sigma_i [-\varepsilon_i.\eta_i^* - x_i^{opt}(\eta_i^*).c_i] \tag{12.16}$$

yields with $\eta_i^{*,opt}=-[c_i^*]^2.[2\varepsilon_i]^{-2}$ an optimal (dual) solution. Thus stability holds for (12.9) at $\eta=0$ and herewith strong duality for (12.9) at $\eta=0$ and (12.16)

It has become apparent throughout the discussions in the previous sections, that the investigations therein must not be claimed to be complete. In particular, for applicability within convex programming it certainly requires several more results. Additionally, to the open questions mentioned already above, we state some more: the requirement of $f_0(\eta,.)$ to be convex likely entail induced constraints for η; such constraints might cause troubles already in modeling convex optimization problems of type (10.1.1). Further, one might think of extending primal-dual algorithms with respect to the dimensions of η^*, intending to accelerate convergence and to increase stability. Such studies likely result in determining saddle points of upper and lower Langrangians in (9.2). This in turn necessitates descriptions of these Lagrangians, whose availability depends mainly on the ability of performing the conjugacy operation explicitly in (12.3). Special cases have to be investigated where the (partial) conjugates of $f_0(\eta,x)$ are as easily obtained as the conjugates of the indicator function. Research activities which aim to answer these open questions are postponed to future work.

Nevertheless, our motivation for studying above duality relations has been lying in the intention to get a condition that ensures subdifferentiability of the optimal value function. Such a result has been achieved with the stability criterion in Theorem 10.2. In the following chapter (III) subdifferentiability, in particular of recourse functions, will assist to ensure existence of extremal bilinear approximates within stochastic two-stage programming.

III Barycentric Approximation

In chapter I we applied the concept of normal integrands that helped to ensure stochastic two-stage problems to be well-defined. With respect to these problems, we further outlined how to raise the saddle property for the integrand and how to model stochastic independence of random data. In this chapter we shall present a **barycentric approximation** technique that exploits these features (saddle property, stochastic independence) and assists in solving stochastic two-stage problems approximately.

Throughout this chapter we base our investigations on stochastic two-stage problems of the form (in accordance with the notations of sections 5,6)

$$\inf_{x\in X} f(x) + \int_{\Omega^{o,1}\times\Omega^{r,1}} \cdots \int_{\Omega^{o,I}\times\Omega^{r,I}} \rho(\eta^1,\xi^1,\ldots,\eta^I,\xi^I,x)\, dP^I(\eta^I,\xi^I)\ldots dP^1(\eta^1,\xi^1)\,. \qquad \text{(III.1.1)}$$

(η^i,ξ^i) represent random vectors with probability measure P^i on $(\Omega^{o,i}\times\Omega^{r,i},\mathcal{B}^{K_i+L_i})$ $(i=1,\ldots,I)$ that are stochastically independent. Sometimes it is useful - for the ease of exposition - to neglect formally the decomposition of (η,ξ) into its stochastic independent subvectors; in this case we then consider (without causing any confusion)

$$\inf_{x\in X} f(x) + \int_{\Omega^o\times\Omega^r} \rho(\eta,\xi,x)\, dP(\eta,\xi)\,. \qquad \text{(III.1.2)}$$

For solving (III.1) we have to take care of the following elements (illustrated in Fig. III.1) : Probability measure P, $\rho(\eta,\xi,x)$ as a parametric (in x) saddle function and as a convex function being parametric in (η,ξ) ; both defining an integral that has to be computed at least approximately. The sum of the obtained expectation functional and the deterministic function, both being defined on the deterministic closed set X, are finally to minimize with respect to x. Methods developed in literature so far for solving stochastic two stage problems of type (III.1) differ mostly in which way and with which priority the above-illustrated elements are handled. According to their characteristics we may classify the methods in 3 groups:

- Approximation Schemes
- Stochastic Approximations
- Mathematical Programming Techniques

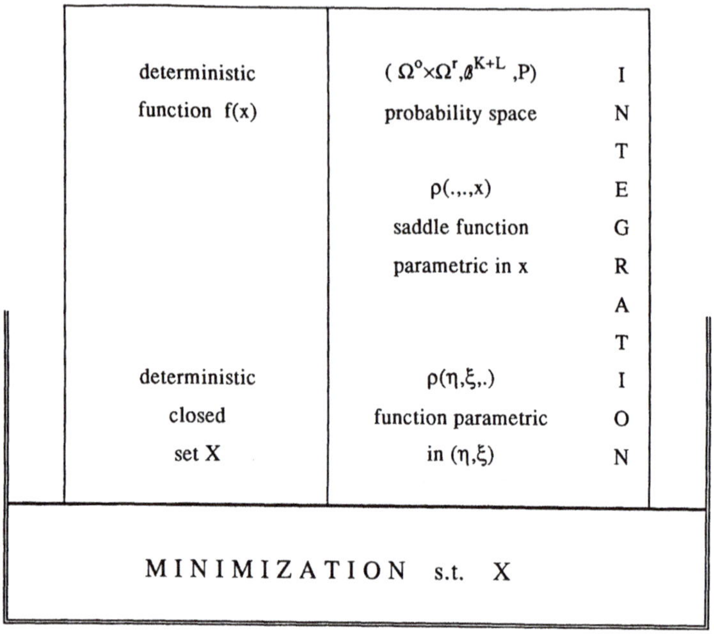

Fig. III.1

Approximation Schemes are characterized by approximating the integrand $\rho(\eta,\xi,x)$ or the probability P measure, finally defining approximate problems for (III.1). Under certain convergence properties one may conclude convergence of the approximate optimal solutions and values to the original.
Stochastic Approximations base on the socalled stochastic (quasi-)gradients, being evaluated with respect to sample information of random data. These stochastic (quasi-)gradients are used for generating feasible solutions and estimates for the corresponding objective value. Under appropriate assumptions the achieved sequences of feasible solutions and objective values converge in a probabilistic sense to the optimal solutions and values of the given problem.
Methods applying *Mathematical Programming Techniques* aim at solving the deterministic problem of type (1.3) associated with (III.1); up to now, such methods are applicable only to highly structured classes of stochastic two-stage problems, like the simple recourse as in (7.4).

The main representatives of these classes will be illustratively outlined in chapter IV.

In this chapter we concentrate on presenting an approach for solving (III.1) that belongs to the class of approximation schemes. We base our investigations on approximating both the integrand as well as the probability measure in an extremal sense : the set of minorants (majorants) for the integrand $\rho(.,.,x)$ may be represented as a feasibility set of a semiinfinite program that aims to determine that minorant (majorant) being maximal (minimal) in expectation with respect to the given probability measure. The associated dual, a generalized moment problem seeks to find that probability measure which coincides in some moments with P and minimizes (maximizes) the expectation of $\rho(.,.,x)$. This points to solving semiinfinite programs and their duals, moment problems, whose optimal solutions are finally in a certain sense best approximates for the integrand as well as for the probability measure, justifying to speak in both cases of extremal approximations. From the computational point of view these approximates have a very nice feature: namely, it will turn out that, with respect to a collection of $\times$-**simplices**, both approximates are completely determined by distinguished cross moments associated with the probability measure involved. This implies that the optimal solutions of these semiinfinite programs, the associated generalized moment problems respectively, will be explicitly available through cross moments. Thus, the task of determining the solutions of semiinfinite programs, moment problems respectively, will reduce to the evaluation of cross moments of P. Further, we shall demonstrate that the evaluations will reduce to those cross moments that refers to the integration of in-(η^i,ξ^i)-bilinear functions, and hence to at most two-dimensional integrals. Finally, as a consequence of the dualization, we may study the goodness of approximates by choice, either with respect to the integrand or with respect to the probability measure.

The motivation for calling the presented approach as **barycentric approximation** will become apparent in section 15 where we present a geometric interpretation of the approximation; sofar, we mention one characteristic: instead of working with the random vectors ξ and η explicitly inside a $\times$-simplex, we shall work with their generalized barycentric weights given with respect to the $\times$-simplex through projections of ξ and η onto the associated simplices. For the ease of understanding we shall not treat the generalized barycentric weights as (new) random vectors, but as multilinear functions.

This chapter starts (section 13) with introducing well-known approximates for convex functions over rectangles and simplices, formulae for their expectations, and extremal measures with their induced inequalities. Subsequently (section 14), we derive multilinear minorants and majorants for saddle functions defined on a

$\times$-simplex. These allow to evaluate (primally) bounds for the expectation of saddle functions, to determine (dually) extremal measures and to state induced inequalities. Section 15 states examples and outlines the geometric interpretation of our approach. In section 16 we introduce 'iterated approximation' with respect to $\times$-simplices that are compatible with the stochastic independence of the underlying probability space. This fact strongly simplifies the evaluations of approximate integrals. We shall proceed with a discussion of refining the approximations appropriately with respect to a nested sequence of $\times$-simplicial partitions. In section 17 we specify the frame under which the achieved results are applied to stochastic two-stage problems. In the subsequent section 18 we prove epi-convergence of the sequence of approximate expectation functionals. Section 19 provides criteria that accelerate convergence, essentially based on studying duality gaps by means of primal and dual optimal solutions. We end this chapter with an iterative completion (section 20) of the barycentric approximation scheme for solving stochastic two-stage problems .

In this section we recall the classical Jensen Inequality (Jensen 1906 [64]) and Edmundson-Madansky Inequality (Madansky 1959 [85]) with their extensions and well-known relations to moment problems. Both have been widely used within the field of stochastic programming (see e.g. Madansky 1960 [86], Avriel and Williams 1970 [2], Ben-Tal and Hochman 1972 [7], Huang, Ziemba and Ben-Tal 1977 [62], Kall and Stoyan 1982 [79]).

Throughout this section, X is an arbitrary set in $\mathbf{R}^n$, P a probability measure on $(\Xi,\mathcal{B}^L)$ for some convex, closed $\Xi \subset \mathbf{R}^L$; for every x in X we define with $\mathcal{F}(x)=\{\rho(.,x):\Xi\to\mathbf{R} \mid \rho(.,x) \text{ finite convex}\}$ a collection of functions, supposed to be integrable with respect to P. The subset of continuous elements of $\mathcal{F}(x)$ let be denoted $\mathcal{F}^c(x)$.

We aim at approximating the expectation functional

$$\int_{\Xi} \rho(\xi,x)\, dP(\xi) \tag{13.1}$$

via the solutions of semiinfinite programs and the associated dual, the moment problems. (For generalized moment problems investigated within probability theory we refer to Krein and Nudel'man 1977 [83]).

At first, let us define a partial ordering, denoted as $\leq^{(c)}$, for probability measures on $(\Xi,\mathcal{B}^L)$ with respect to the set of convex functions, continuous relative to Ξ :

$$P_1 \leq^{(c)} P_2 \;\Leftrightarrow\; \int_{\Xi} \rho(\xi,x)dP_1(\xi) \leq \int_{\Xi} \rho(\xi,x)dP_2(\xi) \quad \forall\, \rho(.,x) \in \mathcal{F}^c(x)\, ,\ x\in X \tag{13.2.1}$$

The set of extremal measures with respect to a set Δ of probability measures are denoted with $\inf^{(c)} \Delta$, $\sup^{(c)} \Delta$ respectively, and defined as

$$\inf^{(c)} \Delta := \{Q_l \mid Q_l \leq^{(c)} Q \;\; , Q_l \in \Delta\, , \forall Q\in \Delta \}, \tag{13.2.2}$$
$$\sup^{(c)} \Delta := \{Q_u \mid Q \leq^{(c)} Q_u \;\; , Q_u \in \Delta\, , \forall Q\in \Delta \}. \tag{13.2.3}$$

Note, '(c)' as superscript points to the class of convex (finite, continuous) functions, based on which extremality is defined. Elements of $\inf^{(c)} \Delta$, $\sup^{(c)} \Delta$ we call **lower extremal probability measures, upper extremal probability**

measures respectively (For the theory of partial orderings of distributions, we refer to Stoyan 1983 [136]).

Remark 13.1: Suppose $P \in \Delta$, $P_l \in \inf^{(c)}\Delta$, $P_u \in \sup^{(c)}\Delta$ and denote $(E\rho)(x)$, $(E_l\,\rho)(x)$, $(E_u\,\rho)(x)$ the expectation functionals of $\rho(\xi,x)$ with respect to P, P_l, P_u. Then, one may conclude that $(E_l\,\rho)(x) \le (E\rho)(x) \le (E_u\,\rho)(x) \quad \forall x \in X$. Moreover, if $\rho(\xi,x)$ is convex on (convex) X for all ξ in Ξ, then we have convexity of the expectation functionals on X, too (Proposition 1.1). This in fact is the motivation for aiming to elements of sets $\inf^{(c)}\Delta$ and $\sup^{(c)}\Delta$.

Remark 13.2: The investigation of extremal measures within stochastic programming starts with Dupacová 1966 [33]. Therein the socalled minimax approach has been introduced, being applicable in case P is not completely known; instead one assumes to know only the set of probability measures to which P belongs (see also Dupacová 1976 [34], 1980 [35], 1987 [36]).

For **any convex** Ξ Jensen Inequality is stated as

$$\rho(\xi^e,x) \le \int_\Xi \rho(\xi,x)dP(\xi) \tag{13.3}$$

where ξ^e denotes the expectation of ξ with respect to P (supposed to exist). (13.3) illustrates that for obtaining a lower bound for the expectation of $\rho(\xi,x)$ we have to evaluate $\rho(.,x)$ at ξ^e. For that which follows in section 14, it will be helpful to outline (obvious) primal and dual details entailed by Jensen Inequality.

Defining P_l as the one-point probability measure on ξ^e,

$$P_l\,[\xi^e] := 1 \tag{13.4}$$

we may write for any convex function $\rho(.,x)$ on Ξ

$$\int_\Xi \rho(\xi,x)dP_l(\xi) \le \int_\Xi \rho(\xi,x)dP(\xi)\,. \tag{13.5}$$

Convexity on Ξ allows to approximate $\rho(.,x)$ from below by a linear affine function $L(.,x) : \Xi \to \mathbf{R}$

$$L(\xi,x) := \langle\alpha(x),\xi\rangle - \beta(x) \le \rho(\xi,x) \tag{13.6}$$

where $\alpha(x) \in \mathbf{R}^L$, $\beta(x) \in \mathbf{R}$ with $\alpha_l(x)$, $l=1,...,L$ denote the coefficients of $L(.,x)$ and $\beta(x)$ the intersect with the ordinate; obviously both depend on x. Let the set of linear affine functions, satisfying (13.6), be denoted $\mathcal{L}(x)$ for fixed x. Integrating (13.6) we obviously obtain a lower approximate for $(E\rho)(x)$ with

$$(EL)(x) := \int_\Xi L(\xi,x)dP(\xi) \leq \int_\Xi \rho(\xi,x)dP(\xi) =: (E\rho)(x). \tag{13.7}$$

For finding the best lower approximate with respect to $\mathcal{L}(x)$, we have to solve:

$$\begin{array}{lcl} \sup \int_\Xi L(\xi,x)dP(\xi) & & \sup \int_\Xi [<\alpha(x),\xi> - \beta(x)]dP(\xi) \\ & \Leftrightarrow & \\ \text{s.t. } L(\xi,x) \in \mathcal{L}(x) & & \text{s.t. } <\alpha(x),\xi> - \beta(x) \leq \rho(\xi,x) \quad \forall \xi \in \Xi . \end{array} \tag{13.8}$$

Due to the linearity of $L(.,x)$ we get

$$\int_\Xi [<\alpha(x),\xi> - \beta(x)] \, dP(\xi) = <\alpha(x),\xi^e> - \beta(x) \tag{13.9}$$

and rewrite (13.8) as semiinfinite program

$$\begin{array}{l} \sup \; <\alpha(x),\xi^e> - \beta(x) \\ \text{s.t. } <\alpha(x),\xi> - \beta(x) \leq \rho(\xi,x) \;\; \forall \, \xi \in \Xi . \end{array} \tag{13.10}$$

The associated dual, the moment problem reads as :

$$\inf \{ \int_\Xi \rho(\xi,x)dQ(\xi) \mid \int_\Xi dQ(\xi)=1 , \int_\Xi \xi dQ(\xi)=\xi^e \} . \tag{13.11}$$

From the semiinfinite program (13.10) we immediately realize that the set of constraints is active at least for $\xi=\xi^e$ if $cl_\xi\rho(\xi^e,x)=\rho(\xi^e,x)$. This is due to maximizing the value of the linear affine function at ξ^e. Hence the optimal objective value is equal $\rho(\xi^e,x)$ as long as $\rho(.,x)$ is supportable from below by a linear function at ξ^e with respect to Ξ. If there exists no linear supporting function for $\rho(.,x)$ at ξ^e (either if $\rho(.,x)$ is only upper semicontinuous at ξ^e or if $\rho(.,x)$ is nonconvex), we would have to accept a duality gap. More than one constraint will be active only in the special case where $\rho(.,x)$ is linear in a neighborhood of ξ^e .

Considering its dual, the moment problem (13.11), we have to minimize the expectation of $\rho(\xi,x)$ with respect to all measures that distribute mass 1 on Ξ and coincide in expectation with ξ^e. The optimal solution of (13.11) is given (with exception of the above-mentioned special cases) by the one-point distribution. For nonconvexity one easily creates examples to become better off with distributing also nonzero masses on points different to ξ^e. Defining Δ as the set of those probability measures that coincide in expectation,

$$\Delta := \{Q \mid \int_\Xi dQ(\xi) = 1 \; ; \; \int_\Xi \xi dQ(\xi) = \xi^e \;\} , \tag{13.12}$$

and employing the above well-known and obvious observations, we state

Proposition 13.1: In the case that $\rho(.,x)$ allows a lower supporting linear function at ξ^e, Jensen Inequality is valid. Further, $P_1 \in \inf^{(c)} \Delta$.

The classical inequality that provides an upper bound for the expectation of convex functions is called Edmundson-Madansky, introduced in Madansky 1959 [85]. This inequality has been originally developed for convex functions of a multivariate random vector whose components have been assumed to be stochastically independent and within a multidimensional rectangle. In recalling this inequality we will concentrate on the approach having been used by the author in [46] to cover simultaneously its generalization for the case that stochastically dependent random components are involved.

Let Ξ **represent a rectangle** of the form

$$\Xi := \times_l \, [b_{l,0} \, , \, b_{l,1} \,] \quad , \tag{13.13}$$

with $b_{l,0} < b_{l,1}$ for $l=1,\ldots,L$. For the ease of exposition we confine to the following notation whenever we work with rectangles: $b_\mu = (b_{1,\mu_1}, \ldots, b_{L,\mu_L})$ where $\mu_l \in \{0,1\}$ for $l=1,\ldots,L$, $b_{\mu'}$ is defined to be the vertex of Ξ being opposite to b_μ, i.e. $\mu_l' = 1-\mu_l$, $l=1,\ldots,L$; writing $\forall\mu$ we implicitly think of all 2^L vertices of Ξ; the set of subsets of $\{1,\ldots,L\}$ will be denoted $\mathcal{M}_\Xi$ and with Λ its element; $\Lambda' := \{1,\ldots,L\} \setminus \Lambda$; further the multilinear functions (i.e. functions linear in its components separately) being of the form

$$(\xi_1,\ldots,\xi_L) \rightarrow \Pi_{l \in \Lambda} \; \xi_l \tag{13.14}$$

will play a central role below, whereby for $\Lambda=\emptyset$ we define $\Pi_{l\in\Lambda}\xi_l := 1$; finally the constant terms $\delta_\Lambda(\mu)$ and V_Ξ are given with

$$\delta_\Lambda(\mu) := \Pi_{l\in\Lambda} (-1)^{(1-\mu_l)} , \quad V_\Xi := \Pi_l (b_{l,1} - b_{l,0}) . \tag{13.15}$$

We will concentrate on multilinear functions being given in general form as:

$$(\xi_1 ,\dots,\xi_L) \rightarrow \Sigma_{\Lambda\in \mathcal{M}_\Xi} c_\Lambda [\Pi_{l\in\Lambda} \xi_l] , \tag{13.16}$$

where $\{c_\Lambda \in \mathbf{R} \mid \forall \Lambda\in\mathcal{M}_\Xi \}$ are the coefficients. We denote with $\mathcal{U}(x)$ the set of upper approximates $U(\xi,x)$ for $\rho(\xi,x)$ being multilinear in ξ and satisfying

$$U(\xi,x) := \Sigma_{\Lambda\in \mathcal{M}_\Xi} c_\Lambda(x) [\Pi_{l\in\Lambda} \xi_l] \geq \rho(\xi,x) \qquad \forall \xi\in\Xi . \tag{13.17}$$

We emphasize the dependence on x with writing $c_\Lambda(x)$. Integrating (13.17) with respect to P yields an upper bound for the expectation of $\rho(.,x)$,

$$(EU)(x) := \int_\Xi \Sigma_{\Lambda\in \mathcal{M}_\Xi} [c_\Lambda(x) [\Pi_{l\in\Lambda} \xi_l]] dP \geq \int_\Xi \rho(\xi,x) dP = (E\rho)(x). \tag{13.18}$$

To obtain the best upper bound with respect to the underlying family $\mathcal{U}(x)$ of upper approximates, we solve the semiinfinite program for fixed x :

$$\begin{array}{lcl} \inf \int_\Xi U(\xi,x)dP(\xi) & & \inf \int_\Xi [\Sigma_{\Lambda\in \mathcal{M}_\Xi} c_\Lambda(x) [\Pi_{l\in\Lambda} \xi_l]] dP(\xi) \\ & \Leftrightarrow & \\ \text{s.t. : } U(.,x) \in \mathcal{U}(x) & & \text{s.t.: } \Sigma_{\Lambda\in \mathcal{M}_\Xi} c_\Lambda(x) [\Pi_{l\in\Lambda}\xi_l] \geq \rho(\xi,x) \ \ \forall \xi\in\Xi . \end{array} \tag{13.19}$$

Defining the cross moments m_Λ as

$$m_\Lambda := \int_\Xi [\Pi_{l\in\Lambda} \xi_l] dP(\xi) , \tag{13.20}$$

which certainly exist due to boundedness of the support, we rewrite (13.19) as

$$\begin{array}{l} \inf \ \Sigma_{\Lambda\in \mathcal{M}_\Xi} c_\Lambda(x) . m_\Lambda \\ \text{s.t.: } \Sigma_{\Lambda\in \mathcal{M}_\Xi} [c_\Lambda(x) [\Pi_{l\in\Lambda} \xi_l]] \geq \rho(\xi,x) \ \ \forall \xi\in\Xi . \end{array} \tag{13.21}$$

Observe that in (13.19) and (13.21) we minimize with respect to $\{c_\Lambda(x), \forall \Lambda \in \mathcal{M}_\Xi\}$. Denoting with $U_{opt}(.,x)$ the optimal solution of (13.19) and applying relation (2.15) in Frauendorfer 1988 [46, p.382] we immediately get (after obvious transformations):

Proposition 13.2 : $U_{opt}(\xi,x)$ being completely determined with

$$c_\Lambda(x) = V_\Xi^{-1} \Sigma_\mu \, \rho(b_\mu,x) \, [\delta_{\Lambda'}(\mu').\delta_\Lambda(\mu)] \, [\, \Pi_{l\in\Lambda'} \, b_{l,\mu'_l} \,] \quad \forall \, \Lambda \in \mathcal{M}_\Xi \,, \qquad (13.22)$$

solves the semiinfinite program (13.21) with $U_{opt}(b_\mu,x) = \rho(b_\mu,x)$ for all μ.

The moment problem associated with (13.21) reads as follows:

$$\sup \{ \int_\Xi \rho(\xi,x) \, dQ \mid \int_\Xi [\, \Pi_{l\in\Lambda} \, \xi_l \,] \, dQ = m_\Lambda \,, \, \forall \, \Lambda \in \mathcal{M}_\Xi \} \qquad (13.23) \,.$$

Due to Theorem 1 in [46] we obtain

Proposition 13.3 : The discrete probability measure P_u being given through

$$P_u \, [b_\mu] = V_\Xi^{-1} \, . \, \Sigma_{\Lambda\in\mathcal{M}_\Xi} \, [\delta_{\Lambda'}(\mu').\delta_\Lambda(\mu)][\, \Pi_{l\in\Lambda'} \, b_{l,\mu'_l} \,] \, m_\Lambda \qquad (13.24)$$

solves the moment problem (13.23).

Obviously, strong duality for (13.19) and (13.23) holds with

$$\int_\Xi U_{opt}(\xi,x) \, dP(\xi) = \int_\Xi \rho(\xi,x) \, dP_u(\xi). \qquad (13.25)$$

Finally, defining the set Δ of probability measures that coincide in all cross moments,

$$\Delta := \{Q \mid \int_\Xi [\, \Pi_{l\in\Lambda} \, \xi_l \,] \, dQ = m_\Lambda \,, \, \forall \, \Lambda \in \mathcal{M}_\Xi \} \qquad (13.26)$$

we restate that $P_u \in \sup^{(c)} \Delta$ [46, p.382].

Substituting $U_{opt}(\xi,x)$ for $U(\xi,x)$ in (13.18) we obtain the generalization of the Edmundson-Madansky Inequality according to

$$\int_{\Xi} \rho(\xi,x)\, dP_u \leq \sum_{\mu} \rho(b_{\mu},x) \left\{ V_{\Xi}^{-1} . \sum_{\Lambda\in \mathfrak{M}_{\Xi}} [\delta_{\Lambda'}(\mu').\delta_{\Lambda}(\mu)][\Pi_{l\in\Lambda'} b_{l,\mu'_l}]\, m_{\Lambda} \right\}. \tag{13.27}$$

Remark 13.3: Edmundson-Madansky Inequality (in its original form for stochastic independent components of ξ) reduces to

$$\int_{\Xi} \rho(\xi,x)\, dP_u(\xi) \leq \sum_{\mu} \rho(b_{\mu},x) \left\{ V_{\Xi}^{-1} . \Pi_l \mid \xi_l^e - b_{l,\mu'_l} \mid \right\} \tag{13.28}$$

with using $m_{\Lambda}=\Pi_{l\in\Lambda}\, \xi_l^e \quad \forall \Lambda\in \mathfrak{M}_{\Xi}$.

Remark 13.4: Kall 1987 [70] proved that $\sup^{(c)} \Delta$ consists of a single element and further illustrated the relation of (13.23) to the univariate moment problem.

Remark 13.5: In employing multilinear functions as upper approximates for $\rho(\xi,x)$ it becomes evident that already $\rho(\xi,x)$ being convex in ξ_l, $l=1,\dots,L$ separately, it suffices to apply the above stated results.

Contrary to Jensen Inequality, applying Edmundson-Madansky Inequality becomes cumbersome if L becomes large, say greater than 10. This is due to the exponentially increasing number of vertices and moments (with respect to L) that have to be handled in (13.24). As an alternative to avoid such effort one can rely on using simplices instead of rectangles, as mentioned in Birge and Wets 1986 [14, p.77]. In case of Ξ being a regular simplex we have to deal with L+1 vertices. Now, as the values of $\rho(.,x)$ at the vertices of Ξ define uniquely a linear function, things become a lot easier in this instance. For easing the exposition of how to majorize, minorize respectively, saddle functions over $\times$-simplices, we will explicitly outline the (rather obvious) inequalities and extremal measures corresponding to simplices.

Let Ξ **respresent a simplex** (regular,closed) in $\mathbf{R}^L$ and its vertices be denoted b_{μ} , $\mu=0,\dots,L$. We define with $\mathcal{U}(x)$ the set of linear affine functions $U(.,x)$ that approximate $\rho(.,x)$ on Ξ from above :

$$U(\xi,x) := \langle\alpha(x),\xi\rangle - \beta(x) \geq \rho(\xi,x) \qquad \forall\xi\in\Xi \tag{13.29}$$

where again $\alpha(x) \in \mathbf{R}^L$, $\beta(x) \in \mathbf{R}$. Integrating (13.29) we trivially obtain an upper bound for $(E\rho)(x)$:

$$(EU)(x) := \int_{\Xi} U(\xi,x)dP(\xi) \geq \int_{\Xi} \rho(\xi,x)dP(\xi) := (E\rho)(x). \tag{13.30}$$

The best upper bound with respect to the set $\mathcal{U}(x)$ is given for fixed x in X :

$$\begin{array}{lcl} \inf \int_{\Xi} U(\xi,x)dP(\xi) & & \inf \int_{\Xi} [\langle\alpha(x),\xi\rangle - \beta(x)]\, dP(\xi) \\ & \Leftrightarrow & \\ \text{s.t.: } U(\xi,x) \in \mathcal{U}(x) & & \text{s.t.: } \langle\alpha(x),\xi\rangle - \beta(x) \geq \rho(\xi,x) \quad \forall \xi \in \Xi \end{array} \tag{13.31}$$

yielding

$$\begin{array}{l} \inf \ \langle\alpha(x),\xi^e\rangle - \beta(x) \\ \text{s.t.: } \langle\alpha(x),\xi\rangle - \beta(x) \geq \rho(\xi,x) \quad \forall \xi \in \Xi. \end{array} \tag{13.32}$$

The way how to derive a solution of (13.32) is obvious and easily outlined with using barycentric weights $\tau(\xi)=(\tau_0(\xi),...,\tau_L(\xi))$ of ξ with respect to the vertices of Ξ : Let T denote the regular matrix of size $(L+1)\times(L+1)$ induced by the (regular) simplex. For determining $\tau(\xi)$ we simply have to solve

$$T.\tau(\xi) = \begin{pmatrix} 1 \\ \xi \end{pmatrix} \qquad \text{with} \qquad T = \begin{pmatrix} 1 & 1 & \dots & 1 \\ b_0 & b_1 & \dots & b_L \end{pmatrix} \tag{13.33}$$

yielding

$$\tau_\mu(\xi) = \langle [T^{-1}]_{\mu,.}\, , (1,\xi)\rangle \quad \forall \mu = 0,...,L \ . \tag{13.34}$$

Due to the fact that $U_{opt}(\xi,x) := \Sigma_\mu\ \tau_\mu(\xi).\rho(b_\mu,x)$ is linear in ξ and given by the values of $\rho(.,x)$ at the vertices b_μ , we conclude $U_{opt}(\xi,x)$ to be uniquely optimal in (13.32). More precisely : $U_{opt}(\xi,x)$ is given with

$$\alpha_l(x) = \Sigma_\mu\ [T^{-1}]_{\mu,l} \cdot \rho(b_\mu,x) \quad (l=1,...,L) \tag{13.35.1}$$
$$\beta(x) = \Sigma_\mu\ [T^{-1}]_{\mu,0} \cdot \rho(b_\mu,x) \ . \tag{13.35.2}$$

The dual of (13.32), the moment problem reads as :

$$\sup \{ \int_{\Xi} \rho(\xi,x)dQ(\xi) \mid \int_{\Xi} dQ(\xi) = 1 , \int_{\Xi} \xi dQ(\xi)= \xi^e \} . \qquad (13.36)$$

As the barycentric weights add up to one and are nonnegative with respect to $\xi \in \Xi$, we obtain with P_u given through

$$P_u [b_\mu] = <[T^{-1}]_{\mu,.} ,(1,\xi^e)> \qquad (13.37)$$

a discrete distribution. Obviously P_u solves (13.36) due to convexity of $\rho(.,x)$. Moreover, $P_u \in \sup^{(c)}\Delta$, with Δ being defined according to (13.12) and Ξ a (regular) simplex .

Above results base mainly on the classical inequalities of Jensen and Edmundson-Madansky and on rectangles, simplices respectively. Upper bounds on the expected value of convex functions have been developed for arbitrary (unbounded) convex polyhedral sets in Gassman and Ziemba 1986 [53] , and for (unbounded) closed convex sets in Birge and Wets 1987 [15] . Birge and Teboulle 1989 [13] studied upper bounds through employing conjugate and first order information of the convex function involved. Upper bounds for the expectation of convex functions of discrete random variables are investigated in Prékopa 1990 [109], based on Lagrangian polynomials. Dulá 1987 [32] considers upper bounds for expectations of sublinear functions with using first and second moments. This idea has been extended in Kall 1991 [73], investigating bounds for recourse functions that are characterized as the maximum of finitely many linear functions. For a review of upper bounds and the related moment problems within stochastic programming we refer to Kall 1988 [72].

Throughout this section, X is an arbitrary set, P a probability measure on $(\Theta\times\Xi,\mathcal{B}^{K+L})$ and $\Theta\times\Xi \subset \mathbf{R}^{K+L}$, with Θ, Ξ being regular (closed) $\times$-**simplices**. For any x in X let us define $\mathcal{F}(x):=\{\rho(.,.,x):\Theta\times\Xi\rightarrow\mathbf{R} \mid \rho(.,.,x)$ finite, concave-convex $\}$ a collection of saddle functions being integrable with respect to P. Again, the subset of continuous elements of $\mathcal{F}(x)$ let be denoted $\mathcal{F}^c(x)$. We consider the following expectation functional :

$$(E\rho)(x) := \int_{\Theta\times\Xi} \rho(\eta,\xi,x)\, dP(\eta,\xi) \,. \tag{14.1}$$

We shall focus, primally, on deriving lower and upper approximates for the saddle function $(\eta,\xi)\rightarrow\rho(\eta,\xi,x)$ and, dually, on deriving lower and upper extremal probability measures with respect to cross moment information of P. To ease exposition and understanding of the subsequent approximation scheme, we concentrate on the cases where Θ, Ξ represent *rectangles* and *simplices*. Its generalization to arbitrary $\times$-simplices Θ, Ξ is of rather technical nature and provides no more insight.

Similarly to the convex case we introduce a partial ordering, denoted $\leq^{(s)}$, for the set of probability measures on $(\Theta\times\Xi,\mathcal{B}^{K+L})$ with respect to the set of saddle functions continuous relative to $\Theta\times\Xi$:

$$P_1 \leq^{(s)} P_2 \Leftrightarrow \int_{\Theta\times\Xi} \rho(\eta,\xi,x)dP_1 \leq \int_{\Theta\times\Xi} \rho(\eta,\xi,x)dP_2 \quad \forall\, \rho(.,.,x)\in\mathcal{F}^c(x),\ x\in X \,. \tag{14.2.1}$$

The set of extremal probability measures with respect to a given set Δ we define in accordance with (13.2) :

$$\inf{}^{(s)} \Delta := \{Q_l \mid Q_l \leq^{(s)} Q \quad Q_l\in\Delta,\ \ \forall Q\in\Delta\} \tag{14.2.2}$$
$$\sup{}^{(s)} \Delta := \{Q_u \mid Q \leq^{(s)} Q_u \quad Q_u\in\Delta,\ \ \forall Q\in\Delta\}. \tag{14.2.3}$$

We start with deriving inequalities and extremal probability measures for Θ, Ξ **being rectangles**. Concerning the rectangle Ξ let the notation coincide with the corresponding one in the convex, rectangular case of section 13; for Θ we have to add that its vertices will be denoted a_v; in particular for $a_{k,0} < a_{k,1}$

$$\Theta := \times_k [a_{k,0} , a_{k,1}] \tag{14.3}$$

with $a_v=(a_{1,v_1} ,..., a_{K,v_K})$ where $v_k \in \{0,1\}$ for k=1,...,K; in dealing with rectangles $a_{v'}$ is defined to be the vertex of Θ being opposite to a_v, i.e. $v_k' = 1-v_k$; with writing $\forall v$ we implicitly think of all 2^K vertices of Θ; the set of subsets of $\{1,...,K\}$ will be denoted $\mathcal{M}_\Theta$ and with Λ its element; $\Lambda' := \{1,...,K\} \setminus \Lambda$; further we will pay attention to the multilinear functions in η

$$(\eta_1,...,\eta_K) \rightarrow \Pi_{k\in\Lambda} \eta_k \quad . \tag{14.4}$$

For $\Lambda=\emptyset$ we define $\Pi_{k\in\Lambda} \eta_k = 1$; finally $\delta_\Lambda(v)$ and V_Θ are given with :

$$\delta_\Lambda(v) := \Pi_{k\in\Lambda} (-1)^{(1-v_k)} \qquad V_\Theta := \Pi_k (a_{k,1} - a_{k,0}) \quad . \tag{14.5}$$

Considering (for any fixed x) the convex functions $\rho(a_v,.,x)$ $\forall v$ and the concave functions $\rho(.,b_\mu,x)$ $\forall \mu$ we may define the set $D_v(x)$ of linear affine functions $d_v(.,x)$ minorizing $\rho(a_v,.,x)$ and the set $H_\mu(x)$ of linear affine functions $h_\mu(.,x)$ majorizing $\rho(.,b_\mu,x)$:

$$D_v(x)=\{d_v(.,x) : \mathbf{R}^L \rightarrow \mathbf{R} \mid d_v(\xi,x)=<\alpha_v(x),\xi> - \beta_v(x) \leq \rho(a_v,\xi,x) \ \ \forall \xi \in \Xi\} \tag{14.6.1}$$
$$H_\mu(x)=\{h_\mu(.,x): \mathbf{R}^K \rightarrow \mathbf{R} \mid h_\mu(\eta,x)=<\alpha_\mu(x),\eta> - \beta_\mu(x) \geq \rho(\eta,b_\mu,x) \ \ \forall \eta \in \Theta\} \tag{14.6.2}$$

where $\alpha_v(x) \in \mathbf{R}^L$, $\alpha_\mu(x) \in \mathbf{R}^K$ and $\beta_v(x)$, $\beta_\mu(x) \in \mathbf{R}$. With

$$\tau_\mu(\xi) = V_\Xi^{-1} \Pi_l \mid \xi_l - b_{l,\mu'_l} \mid \ , \lambda_v(\eta) = V_\Theta^{-1} \Pi_k \mid \eta_k - a_{k,v'_k} \mid \tag{14.7}$$

we further define for any x in X :

$$\mathcal{L}(x) := \{ L(\eta,\xi,x) \mid L(\eta,\xi,x) = \Sigma_v \lambda_v(\eta).d_v(\xi,x) \ \ \forall (\eta,\xi) \in \Theta\times\Xi ; \tag{14.8.1}$$
$$d_v(\xi,x) \in D_v(x) \text{ and } \lambda_v(\eta) \text{ given with (14.7) } \}$$

$$\mathcal{U}(x) := \{ U(\eta,\xi,x) \mid U(\eta,\xi,x) = \Sigma_\mu \tau_\mu(\xi).h_\mu(\eta,x) \ \ \forall (\eta,\xi) \in \Theta\times\Xi ; \tag{14.8.2}$$
$$h_\mu(\eta,x) \in H_\mu(x) \text{ and } \tau_\mu(\xi) \text{ given with (14.7) } \}.$$

Remark 14.1 : $\lambda_v(\eta)$ may be viewed as generalized barycentric weight of η with respect to a_v as it is the product of the barycentric weights of its components η_k with respect to the one-dimensional simplices $[a_{k,0}, a_{k,1}]$ $\forall k$; an analogous view holds for $\tau_\mu(\xi)$.

Lemma 14.1 : Let $\mathcal{L}(x)$, $\mathcal{U}(x)$ be defined as in (14.8.1),(14.8.2). Then $\forall$ $L(\eta,\xi,x) \in \mathcal{L}(x)$ and $\forall$ $U(\eta,\xi,x) \in \mathcal{U}(x)$ we have: $L(\eta,\xi,x)$ is linear in ξ and linear in η_k (k=1,...,K), separately (i.e. multilinear in η); $U(\eta,\xi,x)$ is linear in η and linear in ξ_l (l=1,...,L), separately (i.e. multilinear in ξ). Moreover,

$$L(\eta,\xi,x) \leq \rho(\eta,\xi,x) \leq U(\eta,\xi,x) \ . \tag{14.9}$$

Proof : Observe that for any $\eta_k \in [a_{k,0}, a_{k,1}]$, k=1,...,K we have

$$\Pi_k \mid \eta_k - a_{k,v'_k} \mid = \Sigma_{\Lambda \in \mathcal{M}_\Theta} [\delta_\Lambda(v).\delta_{\Lambda'}(v') \ . \ \Pi_{k \in \Lambda'} a_{k,v'_k}] \ \Pi_{k \in \Lambda} \eta_k \ , \tag{14.10}$$

and hence

$$\lambda_v(\eta) = \Sigma_{\Lambda \in \mathcal{M}_\Theta} c_{\Lambda,v} \ . \ \Pi_{k \in \Lambda} \eta_k \tag{14.11}$$

with $\forall$ $\Lambda \in \mathcal{M}_\Theta$ and $\forall$ v

$$c_{\Lambda,v} = V_\Theta^{-1} [\delta_\Lambda(v).\delta_{\Lambda'}(v') \ . \ \Pi_{k \in \Lambda'} a_{k,v'_k}] \ . \tag{14.12}$$

Analogously, with $\forall$ $\Lambda \in \mathcal{M}_\Xi$ and $\forall$ μ

$$\tau_\mu(\xi) = \Sigma_{\Lambda \in \mathcal{M}_\Xi} c_{\Lambda,\mu} \ . \ \Pi_{l \in \Lambda} \xi_l \tag{14.13}$$

$$c_{\Lambda,\mu} = V_\Xi^{-1} [\delta_\Lambda(\mu).\delta_{\Lambda'}(\mu') \ . \ \Pi_{l \in \Lambda'} b_{l,\mu'_l}] \ . \tag{14.14}$$

Applying (14.11)-(14.14) L(.,.,x) and U(.,.,x) may be written as:

$$L(\eta,\xi,x) = \Sigma_v \Sigma_l \Sigma_{\Lambda \in \mathcal{M}_\Theta} [c_{\Lambda,v} .\alpha_{v,l}(x) \ \xi_l \ .\Pi_{k \in \Lambda} \eta_k \ ] - \\ - \ \Sigma_v \Sigma_{\Lambda \in \mathcal{M}_\Theta} \beta_v(x) \ c_{\Lambda,v} \ .\Pi_{k \in \Lambda} \eta_k \tag{14.15.1}$$

$$U(\eta,\xi,x) = \Sigma_\mu \Sigma_k \Sigma_{\Lambda \in \mathcal{M}_\Xi} [c_{\Lambda,\mu} .\alpha_{\mu,k}(x) \ \eta_k \ .\Pi_{l \in \Lambda} \xi_l \ ] - \\ - \ \Sigma_\mu \Sigma_{\Lambda \in \mathcal{M}_\Xi} \beta_\mu(x) \ c_{\Lambda,\mu} \ .\Pi_{l \in \Lambda} \xi_l \tag{14.15.2}$$

illustrating that L(.,.,x) is linear in ξ, multilinear in η and that U(.,.,x) is linear in η and multilinear in ξ. Finally, due to the properties of $D_\nu(x)$, $H_\mu(x)$, $\rho(a_\nu,.,x)$, $\rho(\eta,b_\mu,x)$, and as

$$\Sigma_\mu \tau_\mu(\xi).b_\mu = \xi \text{ and } \Sigma_\nu \lambda_\nu(\eta).a_\nu = \eta \tag{14.16}$$

(14.9) is evident. //

Integrating (14.9) with respect to P we get lower and upper approximates for $(E\rho)(x)$:

$$(EL)(x) := \int_{\Theta\times\Xi} L(\eta,\xi,x)dP(\eta,\xi) \leq \int_{\Theta\times\Xi} \rho(\eta,\xi,x)dP(\eta,\xi) := (E\rho)(x) \leq$$

$$\leq \int_{\Theta\times\Xi} U(\eta,\xi,x)dP(\eta,\xi) := (EU)(x) \tag{14.17}$$

Before we deduce extremal elements of $\mathcal{L}(x)$, $\mathcal{U}(x)$ respectively, that provide the best lower and upper bounds, we define distinguished cross moments and state their properties. In particular, we consider the cross moments

$$m_{l,\Lambda} := \int_{\Theta\times\Xi} \xi_l \, \Pi_{k\in\Lambda}\eta_k \, dP(\eta,\xi) \quad l=1,\dots,L, \quad \forall\, \Lambda\in \mathcal{M}_\Theta \tag{14.18.1}$$

$$m_\Lambda := \int_{\Theta\times\Xi} \Pi_{k\in\Lambda}\eta_k \, dP(\eta,\xi) \qquad \forall\, \Lambda\in \mathcal{M}_\Theta \tag{14.18.2}$$

$$m_{k,\Lambda} := \int_{\Theta\times\Xi} \eta_k \, \Pi_{l\in\Lambda}\xi_l \, dP(\eta,\xi) \quad k=1,\dots,K, \quad \forall\, \Lambda\in \mathcal{M}_\Xi \tag{14.18.3}$$

$$m_\Lambda := \int_{\Theta\times\Xi} \Pi_{l\in\Lambda}\xi_l \, dP(\eta,\xi) \qquad \forall\, \Lambda\in \mathcal{M}_\Xi \tag{14.18.4}$$

which exist by compactness of $\Theta\times\Xi$.

Lemma 14.2 : Let $c_{\Lambda,\mu}$ and $c_{\Lambda,\nu}$ be defined as in (14.12) and (14.14). The cross moments of (14.18.1) - (14.18.4) satisfy i)-iii) :

i) $p_\Xi(\mu) := \Sigma_{\Lambda\in\mathcal{M}_\Xi} c_{\Lambda,\mu}.m_\Lambda$ are nonnegative $\forall\mu$ and add up to 1 (14.19.1)

$p_\Theta(\nu) := \Sigma_{\Lambda\in\mathcal{M}_\Theta} c_{\Lambda,\nu}.m_\Lambda$ are nonnegative $\forall\nu$ and add up to 1 (14.19.2)

ii) $p_\Xi(\mu) = 0 \Rightarrow \Sigma_{\Lambda\in\mathfrak{M}_\Xi} c_{\Lambda,\mu} . m_{k,\Lambda} = 0 \quad k=1,\dots,K$ (14.20.1)

$p_\Theta(\nu) = 0 \Rightarrow \Sigma_{\Lambda\in\mathfrak{M}_\Theta} c_{\Lambda,\nu} . m_{l,\Lambda} = 0 \quad l=1,\dots,L$ (14.20.2)

iii) Let $\forall\ \Lambda\in\mathfrak{M}_\Xi$ be $(m_{k,\Lambda};\ k=1,\dots,K)\in\mathbf{R}^K$ and analogously let $\forall\ \Lambda\in\mathfrak{M}_\Theta$ be $(m_{l,\Lambda};\ l=1,\dots,L)\in\mathbf{R}^L$. Then

$$p_\Xi(\mu)>0 \Rightarrow p_\Xi(\mu)^{-1}\ [\Sigma_{\Lambda\in\mathfrak{M}_\Xi} c_{\Lambda,\mu} .(m_{k,\Lambda};k=1,\dots,K)\ ] \in \Theta \tag{14.21.1}$$

$$p_\Theta(\nu)>0 \Rightarrow p_\Theta(\nu)^{-1}\ [\Sigma_{\Lambda\in\mathfrak{M}_\Theta} c_{\Lambda,\nu} .(m_{l,\Lambda};l=1,\dots,L)\ ] \in \Xi\ . \tag{14.21.2}$$

Proof : Due to (14.11)-(14.14) and (14.18) we have

$$\Sigma_{\Lambda\in\mathfrak{M}_\Xi} c_{\Lambda,\mu} . m_\Lambda = \int_{\Theta\times\Xi} \tau_\mu(\xi)\ dP(\eta,\xi) \tag{14.22.1}$$

$$\Sigma_{\Lambda\in\mathfrak{M}_\Theta} c_{\Lambda,\nu} . m_\Lambda = \int_{\Theta\times\Xi} \lambda_\nu(\eta)\ dP(\eta,\xi) \tag{14.22.2}$$

proving i) ; these relations and additionally

$$\Sigma_{\Lambda\in\mathfrak{M}_\Xi} c_{\Lambda,\mu} . m_{k,\Lambda} = \int_{\Theta\times\Xi} \eta_k . \tau_\mu(\xi)\ dP(\eta,\xi) \tag{14.23}$$

$$\Sigma_{\Lambda\in\mathfrak{M}_\Theta} c_{\Lambda,\nu} . m_{l,\Lambda} = \int_{\Theta\times\Xi} \xi_l . \lambda_\nu(\eta)\ dP(\eta,\xi) \tag{14.24}$$

imply ii) and iii) due to boundedness of Θ and Ξ. //

Now we are able to formulate

Theorem 14.1 (*Inequality*) : For x in X let $\mathcal{L}(x)$, $\mathcal{U}(x)$ be defined through (14.8.1) , (14.8.2) and let $\eta_\mu=(\eta_{\mu,1},\dots,\eta_{\mu,K})$, $\xi_\nu=(\xi_{\nu,1},\dots,\xi_{\nu,L})$ be given with:

$$\eta_{\mu,k} := \begin{cases} \text{an arbitrary } \eta_k\in[a_{k,0}\ ,\ a_{k,1}\ ] & \text{if : } p_\Xi(\mu) = 0 \\ p_\Xi(\mu)^{-1}\ [\Sigma_{\Lambda\in\mathfrak{M}_\Xi} c_{\Lambda,\mu} . m_{k,\Lambda}\ ] & \text{else} \end{cases} \tag{14.25.1}$$

$$\xi_{\nu,l} := \begin{cases} \text{an arbitrary } \xi_l\in[b_{l,0}\ ,\ b_{l,1}\ ] & \text{if : } p_\Theta(\nu) = 0 \\ p_\Theta(\nu)^{-1}\ [\Sigma_{\Lambda\in\mathfrak{M}_\Theta} c_{\Lambda,\nu} . m_{l,\Lambda}\ ] & \text{else} \end{cases} \tag{14.25.2}$$

Then for any x in X it holds :

$$\sup_{L(.,.,x)\in\mathcal{L}(x)} \int_{\Theta\times\Xi} L(\eta,\xi,x)dP(\eta,\xi) = \Sigma_\nu \, cl_\xi \, \rho(a_\nu,\xi_\nu,x).p_\Theta(\nu) \le$$

$$\le \int_{\Theta\times\Xi} \rho(\eta,\xi,x)dP(\eta,\xi) \le \tag{14.26}$$

$$\le \Sigma_\mu \, cl_\eta \, \rho(\eta_\mu,b_\mu,x).p_\Xi(\mu) = \inf_{U(.,.,x)\in\mathcal{U}(x)} \int_{\Theta\times\Xi} U(\eta,\xi,x)dP(\eta,\xi)$$

Proof : According to (14.15.1)

$$(EL)(x) = \int_{\Theta\times\Xi} L(\eta,\xi,x)dP(\eta,\xi) = \Sigma_\nu \Sigma_l \Sigma_{\Lambda\in\mathfrak{M}_\Theta} [c_{\Lambda,\nu} .\alpha_{\nu,l}(x) \, m_{l,\Lambda}] -$$

$$- \Sigma_\nu \Sigma_{\Lambda\in\mathfrak{M}_\Theta} \beta_\nu(x) \, c_{\Lambda,\nu} .m_\Lambda \; ; \tag{14.27}$$

those terms in (14.27) for which $\Sigma_{\Lambda\in\mathfrak{M}_\Theta} [c_{\Lambda,\nu}.m_\Lambda]>0$ holds, are rewritten as

$$\{\alpha_{\nu,l}(x) \, [\, \Sigma_{\Lambda\in\mathfrak{M}_\Theta} c_{\Lambda,\nu} .m_{l,\Lambda} \,] \, p_\Theta(\nu)^{-1} - \beta_\nu(x)\} \, p_\Theta(\nu) \tag{14.28}$$

with using (14.19). Hence we get

$$\int_{\Theta\times\Xi} L(\eta,\xi,x)dP(\eta,\xi) = \Sigma_\nu \, [<\alpha_\nu(x) \, , \, \xi_\nu > - \, \beta_\nu(x) \,] \, p_\Theta(\nu) \, . \tag{14.29}$$

Due to construction, taking the supremum subject to the lower approximates is equivalent to taking the supremum with respect to the (lower) linear affine functions $d_\nu(.,x)\in D_\nu(x)$, $\forall\nu$ separately. This fact yields

$$\sup_{L(.,.,x)\in\mathcal{L}(x)} \int_{\Theta\times\Xi} L(\eta,\xi,x)dP(\eta,\xi) = \Sigma_\nu \, cl_\xi \, \rho(a_\nu,\xi_\nu,x).p_\Theta(\nu) \; . \tag{14.30}$$

Analogously, we obtain for (EU)(x)

$$\int_{\Theta\times\Xi} U(\eta,\xi,x)dP(\eta,\xi) = \Sigma_\mu \, [<\alpha_\mu(x) \, , \, \eta_\mu > - \, \beta_\mu(x) \,] \, p_\Xi(\mu) \; , \tag{14.31}$$

with taking the infimum with respect to the (upper) linear functions $h_\mu(.,x)\in H_\mu(x)$, $\forall\mu$ separately, we conclude

$$\inf_{U(.,.,x)\in \mathcal{U}(x)} \int_{\Theta\times\Xi} U(\eta,\xi,x)dP(\eta,\xi) = \Sigma_\mu \, cl_\eta \, \rho(\eta_\mu,b_\mu,x).p_\Xi(\mu) \tag{14.32}$$

which completes the proof. //

Corollary 14.1.1: If $p_\Xi(\mu)$, $p_\Theta(\nu)>0$ the η_μ , ξ_ν are given by

$$\eta_\mu = \Sigma_\nu \, a_\nu .[\int_{\Theta\times\Xi} \tau_\mu(\xi).\lambda_\nu(\eta)dP(\eta,\xi)].[\int_{\Theta\times\Xi} \tau_\mu(\xi)dP(\eta,\xi)]^{-1} \tag{14.33.1}$$

$$\xi_\nu = \Sigma_\mu \, b_\mu .[\int_{\Theta\times\Xi} \tau_\mu(\xi).\lambda_\nu(\eta)dP(\eta,\xi)].[\int_{\Theta\times\Xi} \lambda_\nu(\eta)dP(\eta,\xi)]^{-1} \tag{14.33.2}$$

Proof : Utilizing (14.16) in (14.23)-(14.24) and applying (14.22)-(14.25) implies immediately (14.33) . //

Corollary 14.1.2: If $(\eta,\xi)\rightarrow\rho(\eta,\xi,x)$ is subdifferentiable on $\Theta\times\Xi$, then 'supremum' and 'infimum' in (14.26) are attained; i.e. the semiinfinite programs are solvable.

Proof : Subdifferentiability on $\Theta\times\Xi$ implies that $\rho(\eta,\xi,x)$ is linearly supportable on (η_μ,b_μ) $\forall\mu$, (a_ν,ξ_ν) $\forall\nu$; defining $h_\mu(\eta,x)$ to be the supporting linear function for the concave function $\rho(.,b_\mu,x)$ at η_μ $\forall\mu$, symmetrically defining $d_\nu(\xi,x)$ to be the supporting linear function for the convex functions $\rho(a_\nu,.,x)$ at ξ_ν $\forall\nu$, and applying the technique in (14.8) we obtain those linear and upper approximates that solve the semiinfinite programs

$$\sup_{L(.,.,x)\in \mathcal{L}(x)} \int_{\Theta\times\Xi} L(\eta,\xi,x)dP(\eta,\xi) \tag{14.34.1}$$

$$\inf_{U(.,.,x)\in \mathcal{U}(x)} \int_{\Theta\times\Xi} U(\eta,\xi,x)dP(\eta,\xi) \tag{14.34.2}$$

Obviously, the associated $\alpha_\mu(x)$ $\forall\mu$, $\alpha_\nu(x)$ $\forall\nu$ represent the subgradients of $\rho(\eta,\xi,x)$ at (η_μ,b_μ) $\forall\mu$, at (a_ν,ξ_ν) $\forall\nu$. //

In accordance with the results of Frauendorfer 1988 [47] we may state :

Corollary 14.1.3 (*Stochastic Independence*): If $P=P_\eta \times P_\xi$ where P_η and P_ξ denote the marginal probability distributions of P on $(\Theta,\mathcal{B}^K)$ and $(\Xi,\mathcal{B}^L)$, then $\eta_\mu=\eta^e\ \forall\mu$ and $\xi_\nu=\xi^e\ \forall\nu$; (η^e,ξ^e denote the expectation of η, ξ with respect to P_η ,P_ξ).

Proof: Observe that in the case of stochastic independence, it holds that $m_{k,\Lambda}=\eta_k^e.m_\Lambda$ for $k=1,\dots,K$ $\forall\Lambda\in\mathcal{M}_\Xi$, $m_{l,\Lambda}=\xi_l^e.m_\Lambda$ for $l=1,\dots,L$ $\forall\Lambda\in\mathcal{M}_\Theta$ and using (14.22)-(14.25) with respect to $P_\eta \times P_\xi$ yields the assertion. //

Corollary 14.1.4: If $(\eta,\xi)\to\rho(\eta,\xi,x)$ is linear on $\Theta\times\Xi$, then '=' holds in (14.26) .

Proof : Obvious from the fact that then $\rho(.,.,x) \in \mathcal{L}(x) \cap \mathcal{U}(x)$. //

Corollary 14.1.5: If $(\eta,\xi)\to\rho(\eta,\xi,x)$ is a proper closed saddle function, then all elements of the induced equivalence class have the same lower, upper approximate .

Proof : Due to the fact that the lower closed, upper closed elements of an equivalence class of closed saddle functions are unique and coincide with $cl_\xi\rho(\eta,\xi,x)$, $cl_\eta\rho(\eta,\xi,x)$, the assertion becomes evident from (14.26). //

With Lemma 14.2 and with (14.26) of Theorem 14.1, we derived discrete probability measures P_u, P_l on $(\Theta\times\Xi,\mathcal{B}^{K+L})$

$$P_u\ [(\eta_\mu,b_\mu)] = p_\Xi(\mu) \qquad (14.35.1)$$
$$P_l\ [(a_\nu,\xi_\nu)] = p_\Theta(\nu) \qquad (14.35.2)$$

with supports $\{(\eta_\mu,b_\mu)\ \forall\mu\}$, $\{(a_\nu,\xi_\nu)\ \forall\nu\}$ and associated probabilities $p_\Xi(\mu)$, $p_\Theta(\nu)$. Further, observing that P_u is completely determined by $m_{k,\Lambda}$, m_Λ for $k=1,\dots,K$ $\forall\Lambda\in\mathcal{M}_\Xi$, and analogously P_l being completely determined by $m_{l,\Lambda}$, m_Λ for $l=1,\dots,L$ $\forall\Lambda\in\mathcal{M}_\Theta$, we define the following set of probability measures on $(\Theta\times\Xi,\mathcal{B}^{K+L})$:

$$\Delta_\Xi := \{Q \mid \int_{\Theta\times\Xi} \eta_k \prod_{l\in\Lambda}\xi_l\, dQ = m_{k,\Lambda}\ ,\ \int_{\Theta\times\Xi} \prod_{l\in\Lambda}\xi_l\, dQ = m_\Lambda\ \ \forall\Lambda\in\mathcal{M}_\Xi\ ,\ \forall k\} \qquad (14.36.1)$$

$$\Delta_\Theta := \{Q \mid \int_{\Theta\times\Xi} \xi_l \, \Pi_{k\in\Lambda}\eta_k \, dQ = m_{l,\Lambda} \, , \int_{\Theta\times\Xi} \Pi_{k\in\Lambda}\eta_k \, dQ = m_\Lambda \quad \forall\Lambda\in \mathcal{M}_\Theta \, , \forall l\} \, . \quad (14.36.2)$$

We state now

Theorem 14.2 (*Extremal Measures*): Let Δ_Ξ and Δ_Θ be defined as in (14.36). Then $P_u \in \sup^{(s)}\Delta_\Xi$ and $P_l \in \inf^{(s)}\Delta_\Theta$.

Proof: We first show that $P_u \in \Delta_\Xi$, i.e.

$$\Sigma_\mu \, [\, \Pi_{l\in\Lambda_0} b_{l,\mu_l} \cdot \Sigma_{\Lambda\in\mathcal{M}_\Xi} c_{\Lambda,\mu} \cdot m_\Lambda \,] = m_{\Lambda_0} \qquad \forall\Lambda_0\in \mathcal{M}_\Xi \, , \qquad (14.37.1)$$

$$\Sigma_\mu \, \eta_{\mu,k} \cdot [\, \Pi_{l\in\Lambda_0} b_{l,\mu_l} \cdot \Sigma_{\Lambda\in\mathcal{M}_\Xi} c_{\Lambda,\mu} \cdot m_\Lambda \,] = m_{k,\Lambda_0} \qquad \forall\Lambda_0\in \mathcal{M}_\Xi \, , \forall k \qquad (14.37.2)$$

and $P_l \in \Delta_\Theta$, i.e.

$$\Sigma_\nu \, [\, \Pi_{k\in\Lambda_0} a_{k,\nu_k} \cdot \Sigma_{\Lambda\in\mathcal{M}_\Theta} c_{\Lambda,\nu} \cdot m_\Lambda \,] = m_{\Lambda_0} \qquad \forall\Lambda_0\in \mathcal{M}_\Theta \, , \qquad (14.38.1)$$

$$\Sigma_\nu \, \xi_{\nu,l} \cdot [\, \Pi_{k\in\Lambda_0} a_{k,\nu_k} \cdot \Sigma_{\Lambda\in\mathcal{M}_\Theta} c_{\Lambda,\nu} \cdot m_\Lambda \,] = m_{l,\Lambda_0} \qquad \forall\Lambda_0\in \mathcal{M}_\Theta \, , \forall l \, . \qquad (14.38.2)$$

This can easily be seen by applying (14.22) and (14.33) and integration of

$$\Sigma_\mu \, [\, \Pi_{l\in\Lambda_0} b_{l,\mu_l} \cdot \tau_\mu(\xi) \,] = \Pi_{l\in\Lambda_0} \xi_l \qquad \forall\Lambda_0\in \mathcal{M}_\Xi \, , \qquad (14.39.1)$$

$$\Sigma_\nu \, [\, \Pi_{k\in\Lambda_0} a_{k,\nu_k} \cdot \lambda_\nu(\eta) \,] = \Pi_{k\in\Lambda_0} \eta_k \qquad \forall\Lambda_0\in \mathcal{M}_\Theta \, . \qquad (14.39.2)$$

We next show that P_l, P_u respectively, is an extremal element of Δ_Θ, of Δ_Ξ respectively, with respect to the partial ordering $\leq^{(s)}$ defined in (14.2). Any saddle function, finite and continuous on $\Theta\times\Xi$, is convex-closed and concave-closed on $\Theta\times\Xi$. Together with (14.26) and due to the properties of the support points according to (14.25), we conclude that for any $Q\in\Delta_\Theta$

$$\int_{\Theta\times\Xi} cl_\xi \, \rho(\eta,\xi,x) dP_l(\eta,\xi) = \int_{\Theta\times\Xi} \rho(\eta,\xi,x) dP_l(\eta,\xi) \leq \int_{\Theta\times\Xi} \rho(\eta,\xi,x) dQ(\eta,\xi) \, , \qquad (14.40.1)$$

and similarly for any $Q\in\Delta_\Xi$

$$\int_{\Theta\times\Xi} cl_\eta \, \rho(\eta,\xi,x) dP_u(\eta,\xi) = \int_{\Theta\times\Xi} \rho(\eta,\xi,x) dP_u(\eta,\xi) \geq \int_{\Theta\times\Xi} \rho(\eta,\xi,x) dQ(\eta,\xi), \qquad (14.40.2)$$

illustrating that in fact P_l and P_u solve the moment problems

$$\min_{Q\in\Delta_\Theta} \int_{\Theta\times\Xi} \rho(\eta,\xi,x)dQ(\eta,\xi) = \int_{\Theta\times\Xi} \rho(\eta,\xi,x)dP_l(\eta,\xi) \tag{14.41.1}$$

$$\max_{Q\in\Delta_\Xi} \int_{\Theta\times\Xi} \rho(\eta,\xi,x)dQ(\eta,\xi) = \int_{\Theta\times\Xi} \rho(\eta,\xi,x)dP_u(\eta,\xi) \tag{14.41.2}$$

and therefore $P_u \in \sup^{(s)}\Delta_\Xi$, $P_l \in \inf^{(s)}\Delta_\Theta$ which completes the proof. //

Corollary 14.2.1 (*Strong Duality*) : For any x in X with $(\eta,\xi)\to\rho(\eta,\xi,x)$ being continuous at the support points $\{(\eta_\mu,b_\mu)\ \forall\mu\}$, $\{(a_\nu,\xi_\nu)\ \forall\nu\}$ relative to $\Theta\times\Xi$, we obtain strong duality :

$$\sup_{L(.,.,x)\in\mathcal{L}(x)} \int_{\Theta\times\Xi} L(\eta,\xi,x)dP(\eta,\xi) = \min_{Q\in\Delta_\Theta} \int_{\Theta\times\Xi} \rho(\eta,\xi,x)dQ(\eta,\xi) \tag{14.42.1}$$

$$\inf_{U(.,.,x)\in\mathcal{U}(x)} \int_{\Theta\times\Xi} U(\eta,\xi,x)dP(\eta,\xi) = \max_{Q\in\Delta_\Xi} \int_{\Theta\times\Xi} \rho(\eta,\xi,x)dQ(\eta,\xi) \quad . \tag{14.42.2}$$

Proof: The assertion becomes evident from the proof of Theorem 14.2 and the fact that any saddle function continuous at the support points relative to $\Theta\times\Xi$ is concave closed and convex closed on these support points relative to $\Theta\times\Xi$. //

Corollary 14.2.1 (*Strong Duality*): If supp $P_l \subset \Theta \times \mathrm{int}\Xi$, supp $P_u \subset \mathrm{int}\Theta \times \Xi$, then strong duality (14.42) holds for any saddle function $(\eta,\xi)\to\rho(\eta,\xi,x)$ finite on $\Theta\times\Xi$.

Proof : Becomes evident from the proof of Theorem 14.2 and the fact that any saddle function finite on $\Theta\times\Xi$ is convex-closed on $\Theta \times \mathrm{int}\Xi$, concave-closed on $\mathrm{int}\Theta \times \Xi$, respectively. //

Now, let us pass over to the case **Θ, Ξ being regular simplices**, (w.r.t. $\mathbf{R}^K$, $\mathbf{R}^L$) for which things become a lot more comfortable: the vertices of Θ, Ξ are denoted $a_\nu=(a_{1,\nu},\ldots,a_{K,\nu})$, $b_\mu=(b_{1,\mu},\ldots,b_{L,\mu})$ with $\nu=0,\ldots,K$ and $\mu=0,\ldots,L$. The (classical) barycentric weights of η, ξ with respect to Θ, Ξ are denoted $\lambda(\eta):=(\lambda_0(\eta),\ldots,\lambda_K(\eta))$, $\tau(\xi):=(\tau_0(\xi),\ldots,\tau_L(\xi))$ and are given by the (unique) solutions of

$$S.\lambda(\eta) = \begin{pmatrix} 1 \\ \eta \end{pmatrix} \quad \forall\, \eta \in \Theta \tag{14.43.1}$$

$$T.\tau(\xi) = \begin{pmatrix} 1 \\ \xi \end{pmatrix} \quad \forall\, \xi \in \Xi \tag{14.43.2}$$

where S, T denote the regular matrices of sizes (K+1)×(K+1), (L+1)×(L+1) being induced by Θ, Ξ :

$$S := \begin{pmatrix} 1 & 1 & \dots & 1 \\ a_0 & a_1 & \dots & a_K \end{pmatrix} , \qquad T := \begin{pmatrix} 1 & 1 & \dots & 1 \\ b_0 & b_1 & \dots & b_L \end{pmatrix} . \tag{14.44}$$

Restating (13.34) we obtain explicit representations of the barycentric weights :

$$\lambda(\eta) = S^{-1}\begin{pmatrix} 1 \\ \eta \end{pmatrix} , \qquad \lambda_\nu(\eta) = \langle [S^{-1}]_{\nu,.}, (1,\eta)\rangle \quad \nu=0,\dots,K , \tag{14.45.1}$$

$$\tau(\xi) = T^{-1}\begin{pmatrix} 1 \\ \xi \end{pmatrix} , \qquad \tau_\mu(\xi) = \langle [T^{-1}]_{\mu,.}, (1,\xi)\rangle \quad \mu=0,\dots,L . \tag{14.45.2}$$

We adopt $D_\nu(x)$ of (14.6.1) representing the set of linear affine functions $d_\nu(\xi,x)=\langle\alpha_\nu(x),\xi\rangle-\beta_\nu(x)$ approximating $\rho(a_\nu,.,x)$ from below, and analogously $H_\mu(x)$ of (14.6.2) respresenting the set of linear affine functions $h_\mu(\eta,x)=\langle\alpha_\mu(x),\eta\rangle-\beta_\mu(x)$ approximating $\rho(\eta,b_\mu,x)$ from above.

For reasons of symmetry becoming evident below, we outline lower and upper approximation for $(E\rho)(x)$ separately.

Starting with the *Lower Approximation* we define for any x in X

$$\mathcal{L}(x) := \{L(\eta,\xi,x) \mid L(\eta,\xi,x) = \Sigma_\nu\, \lambda_\nu(\eta).d_\nu(\xi,x) \;\; \forall\, (\eta,\xi)\in\Theta\times\Xi ; \;\; d_\nu(\xi,x)\in D_\nu(x) \text{ and } \lambda_\nu(\eta) \text{ given with (14.45.1)}\} . \tag{14.46}$$

Lemma 14.3: $\forall\, L(.,.,x)\in\mathcal{L}(x)$: the function $(\eta,\xi)\rightarrow L(\eta,\xi,x)$ is bilinear in (η,ξ) on $\Theta\times\Xi$ and for any x in X it holds :

$$L(\eta,\xi,x) \le \rho(\eta,\xi,x) \qquad \forall\, (\eta,\xi) \in \Theta\times\Xi . \tag{14.47}$$

Proof: Bilinearity follows immediately from linearity of the barycentric weights $\lambda_v(\eta)$ in η and from linearity of $d_v(\xi,x)$; finally, due to the properties of $D_v(x)$, $\rho(a_v,.,x)$ and as $\Sigma_v \lambda_v(\eta).a_v = \eta$, inequality (14.47) becomes evident. //

Using the general description of bilinear functions

$$(\eta,\xi) \rightarrow \Sigma_k \Sigma_l c_{k,l}.\eta_k.\xi_l + \Sigma_k c_{k,0}.\eta_k + \Sigma_l c_{0,l}.\xi_l + c_{0,0}, \tag{14.48}$$

we may write $L(.,.,x)$ as

$$L(\eta,\xi,x) = \Sigma_k \Sigma_l c_{k,l}(x).\eta_k.\xi_l + \Sigma_k c_{k,0}(x).\eta_k + \Sigma_l c_{0,l}(x).\xi_l + c_{0,0}(x), \tag{14.49}$$

with emphasizing the dependence on x .

The coefficients are easily computable according to :

$$c_{k,l}(x) = \Sigma_v [S^{-1}]_{v,k} \cdot \alpha_{v,l}(x), \quad c_{k,0}(x) = \Sigma_v [S^{-1}]_{v,k} \cdot \beta_v(x) \tag{14.50.1}$$
$$c_{0,l}(x) = \Sigma_v [S^{-1}]_{v,0} \cdot \alpha_{v,l}(x), \quad c_{0,0}(x) = \Sigma_v [S^{-1}]_{v,0} \cdot \beta_v(x) . \tag{14.50.2}$$

Integrating (14.47) with respect to P yields a lower approximate for $(E\rho)(x)$

$$(EL)(x) = \int_{\Theta\times\Xi} L(\eta,\xi,x)dP(\eta,\xi) \leq \int_{\Theta\times\Xi} \rho(\eta,\xi,x)dP(\eta,\xi) = (E\rho)(x) \tag{14.51}$$

and we realize that $(EL)(x)$ is only dependent on the cross moments

$$m_{k,l} = \int_{\Theta\times\Xi} \eta_k.\xi_l \, dP(\eta,\xi), \quad m_{k,0} = \int_{\Theta\times\Xi} \eta_k \, dP(\eta,\xi), \quad m_{0,l} = \int_{\Theta\times\Xi} \xi_l \, dP(\eta,\xi) \tag{14.52}$$

for $k=1,...,K$ and $l=1,...,L$, which exist by boundedness of $\Theta\times\Xi$.

Remark 14.1 : For the ease of understanding, we refrain herewith from denoting the expectation of η, ξ with η^e, ξ^e.

Lemma 14.4: The cross moments of (14.52) satisfy i)-iii) for $\nu=0,...,K$:

i) $p_\Theta(\nu) = \sum_{k=0}^{K} [S^{-1}]_{\nu,k} \cdot m_{k,0}$ are nonnegative $\forall\nu$ and add up to 1 (14.53.1)

ii) $p_\Theta(\nu) = 0 \Rightarrow \sum_{k=0}^{K} [S^{-1}]_{\nu,k} \cdot m_{k,l} = 0$ for $l=1,...,L$ (14.53.2)

iii) $p_\Theta(\nu) > 0 \Rightarrow \xi_\nu=(\xi_{\nu,1},...,\xi_{\nu,L})\in\Xi$ where

$$\xi_{\nu,l} := [\sum_{k=0}^{K} [S^{-1}]_{\nu,k} \cdot m_{k,l}].[p_\Theta(\nu)]^{-1} \quad \text{for } l=1,...,L \ . \qquad (14.53.3)$$

Proof: i) becomes evident through

$$p_\Theta(\nu) = \int_{\Theta\times\Xi} \lambda_\nu(\eta)dP(\eta,\xi) = \sum_{k=0}^{K} [S^{-1}]_{\nu,k} \cdot m_{k,0} \ . \qquad (14.54)$$

For ii) we simply use

$$\Sigma_k [S^{-1}]_{\nu,k} \cdot m_{k,l} = \int_{\Theta\times\Xi} \lambda_\nu(\eta).\xi_l \, dP(\eta,\xi) \quad l=1,...,L \qquad (14.55)$$

and observing

$$\int_{\Theta\times\Xi} \lambda_\nu(\eta)dP(\eta,\xi) = 0 \Leftrightarrow \lambda_\nu(\eta) = 0 \ \text{P-a.s.} \qquad (14.56)$$

together with boundedness of Ξ yields the assertion ii).

On the other hand to prove iii) we argue

$$\Sigma_\mu \int_{\Theta\times\Xi} \lambda_\nu(\eta).\tau_\mu(\xi) \, dP(\eta,\xi) = \int_{\Theta\times\Xi} \lambda_\nu(\eta)dP(\eta,\xi) = p_\Theta(\nu) \qquad (14.57)$$

yielding that

$$\int_{\Theta\times\Xi} \lambda_\nu(\eta).\xi \, dP(\eta,\xi) = \Sigma_\mu \, b_\mu.[\int_{\Theta\times\Xi} \lambda_\nu(\eta).\tau_\mu(\xi) \, dP(\eta,\xi)].p_\Theta(\nu)^{-1} \in \Xi \qquad (14.58)$$

which completes the proof. //

Theorem 14.3 (*Inequality*): For x in X let $\mathcal{L}(x)$ be defined as in (14.46) and $\xi_v=(\xi_{v,1},\ldots,\xi_{v,L})$ be given for l=0,...,L :

$$\xi_{v,l} := \begin{cases} \xi_l \text{ arbitrary with } \xi \in \Xi & \text{if } p_\Theta(v)=0 \\ [\sum_{k=0}^{K} [S^{-1}]_{v,k} \cdot m_{k,l}].[p_\Theta(v)]^{-1} & \text{if } p_\Theta(v)>0 \end{cases} \tag{14.59}$$

Then for any x in X we have

$$\sup_{L(.,.,x)\in\mathcal{L}(x)} \int_{\Theta\times\Xi} L(\eta,\xi,x)dP(\eta,\xi) = \Sigma_v \, cl_\xi \, \rho(a_v,\xi_v,x).p_\Theta(v) \le \int_{\Theta\times\Xi} \rho(\eta,\xi,x)dP. \tag{14.60}$$

Proof : The technique of proving coincides with the one of Theorem 14.1 ; the only difference is that we are working now with (L+1) vertices and bilinear functions. //

For performing the *Upper Approximation* we simply have to apply the above outlined ideas to $-\rho(\eta,\xi,x)$ due to the symmetry of bilinearity. The corresponding notations and results (counterparts to Lemmata 14.3, 14.4 , Theorem 14.3) are stated without proof .

For x in X let the set of upper approximates be defined according to

$$\mathcal{U}(x) := \{ U(\eta,\xi,x) \mid U(\eta,\xi,x) = \Sigma_\mu \, \tau_\mu(\xi).h_\mu(\eta,x) \;\; \forall\, (\eta,\xi) \in \Theta\times\Xi \, ; \; h_\mu(\eta,x)\in H_\mu(x) \text{ and } \tau_\mu(\xi) \text{ given with (14.45.2) } \}. \tag{14.61}$$

Lemma 14.3': $\forall U(.,.,x)\in\mathcal{U}(x)$: the function $(\eta,\xi)\to U(\eta,\xi,x)$ is bilinear in (η,ξ) on $\Theta\times\Xi$ and for any x in X it holds :

$$\rho(\eta,\xi,x) \le U(\eta,\xi,x) \qquad \forall\, (\eta,\xi) \in \Theta\times\Xi \, . \tag{14.62}$$

Using the general description of bilinear functions as in (14.48) we obtain analogously to (14.50) for the coefficients :

$$c_{k,l}(x) = \Sigma_\mu \, [T^{-1}]_{\mu,l} \cdot \alpha_{\mu,k}(x) \, , \; c_{k,0}(x) = \Sigma_v \, [T^{-1}]_{\mu,l} \cdot \beta_\mu(x) \, , \tag{14.63.1}$$
$$c_{0,l}(x) = \Sigma_v \, [T^{-1}]_{\mu,0} \cdot \alpha_{\mu,k}(x) \, , \; c_{0,0}(x) = \Sigma_v \, [T^{-1}]_{\mu,0} \cdot \beta_\mu(x) \, . \tag{14.63.2}$$

Integrating (14.62) with respect to P yields an upper approximate (EU)(x) for $(E\rho)(x)$, being only dependent on the cross moments $m_{k,l}$, $m_{k,0}$, $m_{0,l}$ for k=1,...,K, l=1,...,L , given in 14.52 :

$$(EU)(x) = \int_{\Theta\times\Xi} U(\eta,\xi,x)dP(\eta,\xi) \geq \int_{\Theta\times\Xi} \rho(\eta,\xi,x)dP(\eta,\xi) = (E\rho)(x) \, . \quad (14.64)$$

Lemma 14.4' : The cross moments of (14.52) satisfy i) - iii) for μ=0,...,L :

i) $p_\Xi(\mu) = \sum_{l=0}^{L} [T^{-1}]_{\mu,l} \cdot m_{0,l}$ are nonnegative $\forall \nu$ and add up to 1 (14.65.1)

ii) $p_\Xi(\mu) = 0 \Rightarrow \sum_{l=0}^{L} [T^{-1}]_{\mu,l} \cdot m_{k,l} = 0$ for k=1,...,K (14.65.2)

iii) $p_\Xi(\mu) > 0 \Rightarrow \eta_\mu=(\eta_{\mu,1},\ldots,\eta_{\mu,K}) \in \Theta$ where

$$\eta_{\mu,k} := [\sum_{l=0}^{L} [T^{-1}]_{\mu,l} \cdot m_{k,l}].[p_\Xi(\mu)]^{-1} \quad \text{for } k=1,\ldots,K \, . \quad (14.65.3)$$

Theorem 14.3' (*Inequality*): For $\mathcal{U}(x)$ being defined as in (14.61) and let $\eta_\mu=(\eta_{\mu,1},\ldots,\eta_{\mu,K})$, be given with (for k=1,...,K):

$$\eta_{\mu,k} := \begin{cases} \eta_k \text{ arbitrary with } \eta \in \Theta & \text{if } p_\Xi(\mu) = 0 \\ [\sum_{l=0}^{L} [T^{-1}]_{\mu,l} \cdot m_{k,l}].[p_\Xi(\mu)]^{-1} & \text{if } p_\Xi(\mu) > 0 \end{cases} \quad (14.66)$$

Then for any x in X we have,

$$\int_{\Theta\times\Xi} \rho(\eta,\xi,x)dP(\eta,\xi) \leq \Sigma_\mu \, cl_\eta \, \rho(\eta_\mu,b_\mu,x).p_\Xi(\mu) = \inf_{U(.,.,x)\in\mathcal{U}(x)} \int_{\Theta\times\Xi} U(\eta,\xi,x)dP. \quad (14.67)$$

The subsequent Corollaries are implied by Theorems 14.3, 14.3' :

Corollary 14.3.1 : If $p_\Xi(\mu), p_\Theta(\nu) > 0$ then η_μ , ξ_ν may be represented as :

$$\eta_\mu = \Sigma_\nu\, a_\nu \,.[\int_{\Theta\times\Xi} \tau_\mu(\xi).\lambda_\nu(\eta)dP(\eta,\xi)].[\int_{\Theta\times\Xi} \tau_\mu(\xi)dP(\eta,\xi)]^{-1} \qquad (14.68.1)$$

$$\xi_\nu = \Sigma_\mu\, b_\mu \,.[\int_{\Theta\times\Xi} \tau_\mu(\xi).\lambda_\nu(\eta)dP(\eta,\xi)].[\int_{\Theta\times\Xi} \lambda_\nu(\eta)dP(\eta,\xi)]^{-1} \,. \qquad (14.68.2)$$

Proof : Evident from the proof of Lemma 14.4 (14.4'). //

Corollary 14.3.2 (*Degenerate Case*) :

$$p_\Xi(\mu')=0 \;\Leftrightarrow\; \text{supp } P \subset \Theta \times \text{co}[\{b_0, \ldots, b_L\}\setminus\{b_{\mu'}\}] \qquad (14.69.1)$$

$$p_\Theta(\nu')=0 \;\Leftrightarrow\; \text{supp } P \subset \text{co}[\{a_0, \ldots, a_K\}\setminus\{a_{\nu'}\}] \times \Xi \,. \qquad (14.69.2)$$

Proof : The assertion becomes obvious with

$$p_\Xi(\mu')= \int_{\Theta\times\Xi} \tau_{\mu'}(\xi)dP(\eta,\xi) = 0 \;\Leftrightarrow\; \tau_{\mu'}(\xi)=0 \quad P\text{ - a.s. }, \qquad (14.70.1)$$

$$p_\Theta(\nu')= \int_{\Theta\times\Xi} \lambda_{\nu'}(\eta)dP(\eta,\xi) = 0 \;\Leftrightarrow\; \lambda_{\nu'}(\eta)=0 \quad P\text{ - a.s. }, \qquad (14.70.2)$$

due to the nonnegativity of the barycentric weights . //

Corollary 14.3.3 : If $\rho(\eta,\xi,x)$ is subdifferentiable on $\Theta\times\Xi$ then 'supremum' and 'infimum' in (14.60), (14.67) are attained; i.e. the corresponding semiinfinite programs are solvable.

Proof : Analogously to Corollary 14.1.2 . //

Corollary 14.3.4 (*Stochastic Independence*): If $P=P_\eta \times P_\xi$ where P_η and P_ξ denote the marginal probability distributions of P on $(\Theta,\mathcal{B}^K)$ and $(\Xi,\mathcal{B}^L)$, then $\eta_\mu=\eta^e$ $(\mu=0,\ldots,L)$ and $\xi_\nu=\xi^e$ $(\nu=0,\ldots,K)$; (η^e,ξ^e denote the expectation of η, ξ with respect to P_η ,P_ξ).

Proof : Observe that in the case of stochastic independence we have $m_{k,l}= \eta_k^{\,e}.\xi_l^{\,e}$ for $k=1,\ldots,K$, $l=1,\ldots,L$ and using (14.53.3), (14.65.3) yields the assertion. //

Corollary 14.3.5: If $\rho(.,.,x)$ is linear on $\Theta\times\Xi$, then (14.60), (14.67) holds with '='.

Proof: Obvious from the fact that $\rho(.,.,x) \in \mathcal{L}(x)\cap\mathcal{U}(x)$. //

Corollary 14.3.6 : If $\rho(.,.,x)$ is closed, then all elements of the induced equivalence class have the same lower, upper respectively, approximate .

Proof: Analogously to Corollary 14.1.5 . //

With Lemma 14.4 (14.4') and with Theorem 14.3 (14.3') we derived discrete probability measures P_u, P_l on $(\Theta\times\Xi,\mathcal{B}^{K+L})$

$$P_l\ [(a_\nu,\xi_\nu)] = p_\Theta(\nu) \quad , \tag{14.71.1}$$
$$P_u\ [(\eta_\mu,b_\mu)] = p_\Xi(\mu) \quad . \tag{14.71.2}$$

with supports $\{(a_\nu,\xi_\nu),\nu=0,...,K\}$, $\{(\eta_\mu,b_\mu),\mu=0,...,L\}$ and associated probabilities $p_\Xi(\mu)$, $p_\Theta(\nu)$. Further, observing that P_l and P_u are completely determined by $m_{\nu,\mu}$, $\nu=0,...,K$, $\mu=0,...,L$, we define in accordance the following set of probability measures :

$$\Delta := \{Q|\int_{\Theta\times\Xi} dQ=1,\ \int_{\Theta\times\Xi} \eta_k\,\xi_l\,dQ=m_{k,l}\,,\ \int_{\Theta\times\Xi} \xi_l\,dQ=m_{0,l}\,,\ \int_{\Theta\times\Xi} \eta_k\,dQ= m_{k,0}\ \forall k,l\}. \tag{14.72}$$

We state now

Theorem 14.4 (*Extremal Measures*) : Let Δ be defined as in (14.72). Then $P_u \in \sup^{(s)}\Delta$ and $P_l \in \inf^{(s)}\Delta$.

Proof: We start with proving $P_u \in \Delta$, i.e.

$$\Sigma_\mu\ [\ b_{l,\mu}\cdot p_\Xi(\mu)] = m_{0,l} \qquad l=1,...,L \tag{14.73.1}$$
$$\Sigma_\mu\ \eta_{\mu,k}\cdot[\ b_{l,\mu}\cdot p_\Xi(\mu)\ ] = m_{k,l} \qquad l=1,...,L \quad k=1,..,K, \tag{14.73.2}$$
$$\Sigma_\mu\ \eta_{\mu,k}\cdot p_\Xi(\mu) = m_{k,0} \qquad k=1,..,K \tag{14.73.3}$$

and $P_l \in \Delta$ i.e.

$$\Sigma_\nu [\, a_{k,\nu} \cdot p_\Theta(\nu)] = m_{k,0} \qquad k=1,..,K \qquad (14.74.1)$$
$$\Sigma_\nu\, \xi_{\nu,l} \cdot [a_{k,\nu} \cdot p_\Theta(\nu)\,] = m_{k,l} \qquad l=1,...,L \quad k=1,..,K \qquad (14.74.2)$$
$$\Sigma_\nu\, \xi_{\nu,l} \cdot p_\Theta(\nu) \;= m_{0,l} \qquad l=1,...,L. \qquad (14.74.3)$$

This can easily be seen by applying (14.53.1),(14.54),(14.65.1) and (14.68.1), (14.68.2) of Corollary 14.3.2 (14.57) and (14.58) and additionally integration of

$$\Sigma_\mu [\, b_{l,\mu} \cdot \tau_\mu(\xi)\,] = \xi_l \qquad (14.75.1)$$
$$\Sigma_\nu [\, a_{k,\nu} \cdot \lambda_\nu(\eta)\,] = \eta_k\,. \qquad (14.75.2)$$

Next we show that P_l and P_u are extremal elements of Δ with respect to the partial ordering $\leq^{(s)}$ of (14.2). Again, for any saddle function finite and continuous on $\Theta\times\Xi$ we apply (14.60), (14.67) and obtain for any $Q\in\Delta$

$$\int_{\Theta\times\Xi} cl_\xi\, \rho(\eta,\xi,x)dP_l(\eta,\xi) = \int_{\Theta\times\Xi} \rho(\eta,\xi,x)dP_l(\eta,\xi) \leq \int_{\Theta\times\Xi} \rho(\eta,\xi,x)dQ(\eta,\xi) \qquad (14.76.1)$$

$$\int_{\Theta\times\Xi} cl_\eta\, \rho(\eta,\xi,x)dP_u(\eta,\xi) = \int_{\Theta\times\Xi} \rho(\eta,\xi,x)dP_u(\eta,\xi) \geq \int_{\Theta\times\Xi} \rho(\eta,\xi,x)dQ(\eta,\xi)\,. \qquad (14.76.2)$$

Hence P_l and P_u solve the moment problems

$$\min_{Q\in\Delta} \int_{\Theta\times\Xi} \rho(\eta,\xi,x)dQ(\eta,\xi) = \int_{\Theta\times\Xi} \rho(\eta,\xi,x)dP_l(\eta,\xi) \qquad (14.77.1)$$

$$\max_{Q\in\Delta} \int_{\Theta\times\Xi} \rho(\eta,\xi,x)dQ(\eta,\xi) = \int_{\Theta\times\Xi} \rho(\eta,\xi,x)dP_u(\eta,\xi) \qquad (14.77.2)$$

illustrating that $P_u \in \sup^{(s)}\Delta$ and $P_l \in \inf^{(s)}\Delta$. //

Corollary 14.4.1 (*Strong Duality*) : For any x in X with $(\eta,\xi)\to\rho(\eta,\xi,x)$ being continuous at the support points $\{(\eta_\mu,b_\mu)\ \forall\mu\}$, $\{(a_\nu,\xi_\nu)\ \forall\nu\}$ relative to $\Theta\times\Xi$, we obtain strong duality :

$$\sup_{L(.,.,x)\in\mathcal{L}(x)} \int_{\Theta\times\Xi} L(\eta,\xi,x)dP(\eta,\xi) = \min_{Q\in\Delta} \int_{\Theta\times\Xi} \rho(\eta,\xi,x)dQ(\eta,\xi) \qquad (14.78.1)$$

$$\inf_{U(.,.,x)\in\mathcal{U}(x)} \int_{\Theta\times\Xi} U(\eta,\xi,x)dP(\eta,\xi) = \max_{Q\in\Delta} \int_{\Theta\times\Xi} \rho(\eta,\xi,x)dQ(\eta,\xi)\;. \qquad (14.78.2)$$

Proof: The assertion becomes evident from the proof of Theorem 14.4 and the fact that any saddle function continuous at support points relative to $\Theta\times\Xi$ is convex- closed and concave-closed at these points. //

Corollary 14.4.1 (*Strong Duality*): If supp $P_l \subset \Theta \times \text{int}\Xi$, supp $P_u \subset \text{int}\Theta \times \Xi$, then strong duality holds for any saddle function $(\eta,\xi) \rightarrow \rho(\eta,\xi,x)$ finite on $\Theta\times\Xi$.

Proof : Becomes evident from the proof of Theorem 14.4 and the fact that any saddle function finite on $\Theta\times\Xi$ is convex-closed on $\Theta \times \text{int}\Xi$, concave-closed on $\text{int}\Theta \times \Xi$, respectively. //

Let us conclude with some remarks on the above outlined approach: we focused on approximating a collection $\mathcal{F}(x)$ of saddle functions and its expectation functional with respect to $\times$-simplices, supposed to cover the support of the random vector. We succeeded in deriving best possible approximates with respect to all multilinear, bilinear respectively, minorants and majorants over $\Theta\times\Xi$, being determined by the 'optimal' support points - (14.25), or (14.59), (14.66) - where any subdifferentiable element of $\mathcal{F}(x)$ is recommended to be supported. In fact, one may trust on any minorant or majorant of that type with or without supporting the saddle function at some points and could therefore avoid the evaluation of the 'optimal' supporting points according to (14.25) or according to (14.59),(14.66). In other words one could rely on any feasible approximate of the associated semiinfinite programs in (14.34), or in (14.78). In that case one should be aware of the fact that the obtained bounds strongly depend on the subgradients with respect to these approximates and may destroy any structural property (like convexity) of the expectation functional (see Frauendorfer 1988 [48]) .
Moreover, we have to stress that in case of simplices, the formulae for extremal probability measures consist of at most 2-dimensional integrals of bilinear terms. This fact strongly simplifies and accelerates their evaluations. Thus, the numerical effort for evaluating extremal probability measures depends strongly on whether we are working with rectangles or with simplices: in the case of rectangles we have exponential increase in the number $(L.2^K+K.2^L)$ of cross moments with respect to the dimension of (η,ξ); to the contrary, working with simplices we have to evaluate (K+1).(L+1) cross moments and therefore polynomially increasing with respect to the dimension of (η,ξ).

Extensions of the above approach may be found in Edirisinghe and Ziemba 1989 [38], where the expectation of saddle functions are investigated with respect to arbitrary, bounded polytopes.

15 Examples and Geometric Interpretation

Below we outline an example and give a geometric interpretation of the primal and dual approximations derived in the preceding section. For geometric interpretation of the purely convex case, mainly from the dual point of view, we refer to Kall *et al.* 1988 [78].

<u>Example 15.1</u> : We take L=K=1 and $\Theta=[a_0,a_1]=[0,1]$ and $\Xi=[b_0,b_1]=[0,1]$. Let (η,ξ) be discretely distributed according to P being given with $P[(a,0)]=1/2$ and $P[(0,b)]=1/2$ where $a\in[a_0,a_1]$ and $b\in[b_0,b_1]$. We shall compute the extremal lower and upper distributions associated with P and $\Theta\times\Xi$.

We start with determining the distributions of $(\lambda_\nu(\eta),\tau_\mu(\xi))$ for $\nu=0,1$ and $\mu=0,1$:

$$
\begin{aligned}
P[(\lambda_0(\eta),\tau_0(\xi))=(1,1)] &= P[(0,0)]=0\\
P[(\lambda_0(\eta),\tau_0(\xi))=(1,1-b)] &= P[(0,b)]=1/2\\
P[(\lambda_0(\eta),\tau_0(\xi))=(1-a,1)] &= P[(a,0)]=1/2\\
P[(\lambda_0(\eta),\tau_0(\xi))=(1-a,1-b)] &= P[(0,b)]=0
\end{aligned}
$$

$$
\begin{aligned}
P[(\lambda_0(\eta),\tau_1(\xi))=(1,0)] &= P[(0,0)]=0\\
P[(\lambda_0(\eta),\tau_1(\xi))=(1,b)] &= P[(0,b)]=1/2\\
P[(\lambda_0(\eta),\tau_1(\xi))=(1-a,0)] &= P[(a,0)]=1/2\\
P[(\lambda_0(\eta),\tau_1(\xi))=(1-a,b)] &= P[(0,b)]=0
\end{aligned}
$$

$$
\begin{aligned}
P[(\lambda_1(\eta),\tau_0(\xi))=(0,1)] &= P[(0,0)]=0\\
P[(\lambda_1(\eta),\tau_0(\xi))=(0,1-b)] &= P[(0,b)]=1/2\\
P[(\lambda_1(\eta),\tau_0(\xi))=(a,1)] &= P[(a,0)]=1/2\\
P[(\lambda_1(\eta),\tau_0(\xi))=(a,1-b)] &= P[(a,b)]=0
\end{aligned}
$$

$$P[(\lambda_1(\eta),\tau_1(\xi))=(0,0)] \quad = \quad P[(0,0)]=0$$
$$P[(\lambda_1(\eta),\tau_1(\xi))=(0,b)] \quad = \quad P[(0,b)]=1/2$$
$$P[(\lambda_1(\eta),\tau_1(\xi))=(a,0)] \quad = \quad P[(a,0)]=1/2$$
$$P[(\lambda_1(\eta),\tau_1(\xi))=(a,b)] \quad = \quad P[(a,b)]=0 \ .$$

This yields for $\int \lambda_\nu(\eta).\tau_\mu(\xi)\, dP(\eta,\xi)$ with $\nu=0,1$ and $\mu=0,1$:

$$\int \lambda_0(\eta).\tau_0(\xi)dP(\eta,\xi) \quad = \quad 1 - (1/2).b - (1/2).a$$
$$\int \lambda_0(\eta).\tau_1(\xi)dP(\eta,\xi) \quad = \quad (1/2).b$$
$$\int \lambda_1(\eta).\tau_0(\xi)\, dP(\eta,\xi) \quad = \quad (1/2).a$$
$$\int \lambda_1(\eta).\tau_1(\xi).\, dP(\eta,\xi) \quad = \quad 0$$

and for $\int \lambda_\nu(\eta).dP(\eta,\xi)$ $(\nu=0,1)$ and $\int \tau_\mu(\xi)\, dP(\eta,\xi)$ $(\mu=0,1)$:

$$\int \lambda_0(\eta).dP(\eta,\xi) \ = \ 1 - (1/2).a$$
$$\int \tau_0(\xi)\, dP(\eta,\xi) \ = \ 1-(1/2).b$$
$$\int \lambda_1(\eta).dP(\eta,\xi) \ = \ (1/2).a$$
$$\int \tau_1(\xi)\, dP(\eta,\xi) \ = \ (1/2).b$$

This allows to apply formulae (14.68.1) and (14.68.2) to determine the atoms (a_0,ξ_0), (a_1,ξ_1) and (η_0,b_0),(η_1,b_1) of the extremal lower and upper distributions. With having $a_0=0$, $a_1=1$, $b_0=0$, $b_1=1$ we get :

$$\xi_0 = b_0 .[\int \lambda_0(\eta).\tau_0(\xi)dP(\eta,\xi)].[\int \lambda_0(\eta)dP(\eta,\xi)\ ]^{-1} \ +$$
$$+ b_1 .[\int \lambda_0(\eta).\tau_1(\xi)dP(\eta,\xi)].[\int \lambda_0(\eta)dP(\eta,\xi)\ ]^{-1} = (1/2).b.[1-(1/2)a]^{-1}$$

$$\xi_1 = b_0 .[\int \lambda_1(\eta).\tau_0(\xi)dP(\eta,\xi)].[\int \lambda_1(\eta)dP(\eta,\xi)\ ]^{-1} \ +$$
$$+ b_1 .[\int \lambda_1(\eta).\tau_1(\xi)dP(\eta,\xi)].[\int \lambda_1(\eta)dP(\eta,\xi)\ ]^{-1} = 0$$

$$\eta_0 = a_0 .[\int \lambda_0(\eta).\tau_0(\xi)dP(\eta,\xi)].[\int \tau_0(\eta)dP(\eta,\xi)]^{-1} +$$
$$+ a_1 .[\int \lambda_1(\eta).\tau_0(\xi)dP(\eta,\xi)].[\int \tau_0(\eta)dP(\eta,\xi)]^{-1} = (1/2).a.[1-(1/2)b]^{-1}$$

$$\eta_1 = a_0 .[\int \lambda_0(\eta).\tau_1(\xi)dP(\eta,\xi)].[\int \tau_1(\eta)dP(\eta,\xi)]^{-1} +$$
$$+ a_1 .[\int \lambda_1(\eta).\tau_1(\xi)dP(\eta,\xi)].[\int \tau_1(\eta)dP(\eta,\xi)]^{-1} = 0 \ .$$

The associated probabilities for (a_0,ξ_0), (a_1,ξ_1) and (η_0,b_0),(η_1,b_1) are given with:

$$P_l\,[(a_0,\xi_0)] = p_\Theta(0) = [\int \lambda_0(\eta)dP(\eta,\xi)] \ = \ 1-(1/2).a$$
$$P_l\,[(a_1,\xi_1)] = p_\Theta(1) = [\int \lambda_1(\eta)dP(\eta,\xi)] \ = \ (1/2).a$$

$$P_u\,[(\eta_0,b_0)] = p_\Xi(0) = [\int \tau_0(\xi).dP(\eta,\xi)] \ = \ 1-(1/2).b$$
$$P_u\,[(\eta_1,b_1)] = p_\Xi(1) = [\int \tau_1(\xi).dP(\eta,\xi)] \ = \ (1/2).b$$

Primal Interpretation

We again take K=L=1 and denote the vertices of the $\times$-simplex with (a_0,b_0), (a_1,b_0), (a_0,b_1) and (a_1,b_1). The atoms of the lower and upper extremal distributions are given with (a_0,ξ_0), (a_1,ξ_1). (η_0,b_0),(η_1,b_1). Below, in Fig.15.1.a we have the bilinear function $L(\eta,\xi,x)$ as lower approximate for the saddle function $\rho(\eta,\xi,x)$ with supporting $\rho(\eta,\xi,x)$ at (a_0,ξ_0), (a_1,ξ_1); apparently, $L(\eta,\xi,x)$ is determined through the affine linear functions that support the convex functions $\rho(a_0,.,x)$ and $\rho(a_1,.,x)$ at ξ_0 and ξ_1. The different slopes of these linear affine functions entail bilinearity of $L(\eta,\xi,x)$.

In Fig. 15.1.b we illustrate the upper approximation: the bilinear function $U(\eta,\xi,x)$ is upper approximate for the saddle function $\rho(\eta,\xi,x)$ with supporting $\rho(\eta,\xi,x)$ at (η_0,b_0), (η_1,b_1); apparently, $U(\eta,\xi,x)$ is determined through the affine linear functions that support the concave functions $\rho(.,b_0,x)$ and $\rho(.,b_1,x)$ at η_0 and η_1. Again, the different slopes of these supporting linear functions entail bilinearity of $U(\eta,\xi,x)$.

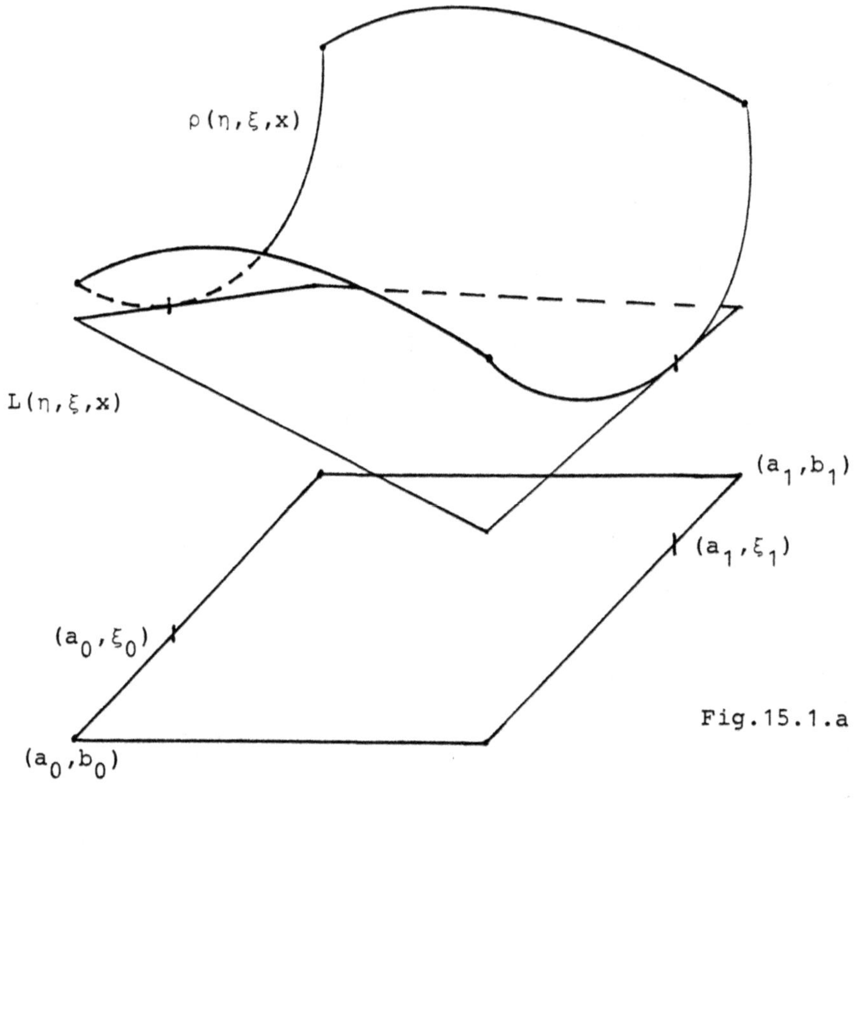

Fig.15.1.a

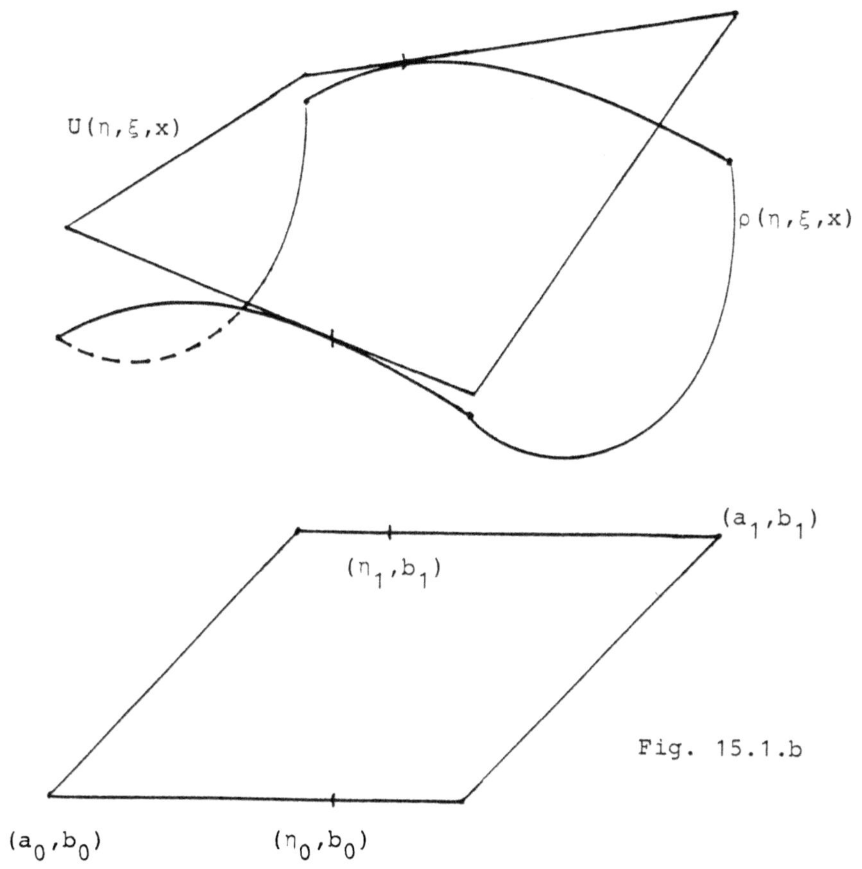

Fig. 15.1.b

We set now K=2 and L=1. Let $\Theta=co\{a_0, a_1, a_2\}$ and $\Xi=co\{b_0, b_1\}$ and the vertices of the $\times$-simplex $\Theta\times\Xi$ be denoted (a_0,b_0), (a_1,b_0), (a_2,b_0), (a_0,b_1), (a_1,b_1) and (a_2,b_1). Let P be a discrete distribution with 3 atoms denoted A, B and C having equal probability of 1/3 (for simplicity).

For illustrating the derivation of the upper extremal distribution let us consider Fig.15.2.a. Projecting A,B, and C onto the 2-dimensional simplex $\Theta\times\{b_0\}$ yields the points A_0, B_0 and C_0. The sizes of the boxes at these projections shall illustrate the mass m(.) that is assigned to these points: in particular, $m(A_0)=(1/3).\tau_0(A)$, $m(B_0)=(1/3).\tau_0(B)$ and $m(C_0)=(1/3).\tau_0(C)$. Apparently, the mass assigned to these projections A_0, B_0 and C_0 is proportional to the barycentric weights of A, B and C with respect to their ξ-components and with respect to the vertex b_0 (note that the projections has been performed onto $\Theta\times\{b_0\}$). So, the 'closer' the atom to $\Theta\times\{b_0\}$, the more mass is assigned; whereby 'closer' is measured with the associated barycentric weight $\tau_0(.)$. The entire mass being distributed onto $\Theta\times\{b_0\}$ is equal to the expectation of $\tau_0(.)$; in our case this is $[\tau_0(A)+\tau_0(B)+\tau_0(C)].(1/3)$ Therefore, we obtained a mass distribution on the simplex $\Theta\times\{b_0\}$ that induces a generalized barycenter, denoted η_0 .
The analogous argumentation may be applied with the projections of A,B, and C onto the 2-dimensional simplex $\Theta\times\{b_1\}$ yielding A_1, B_1 and C_1. The sizes of the boxes at these projections illustrate the mass m(.) that is assigned to these points: in particular, $m(A_1)=(1/3).\tau_1(A)$, $m(B_1)=(1/3).\tau_1(B)$ and $m(C_1)=(1/3).\tau_1(C)$. Apparently , the mass assigned to these projections A_1, B_1 and C_1 is proportional to the barycentric weights of A, B and C with respect to their ξ-components and with respect to the vertex b_1; the 'closer' the atom to $\Theta\times\{b_1\}$, the larger $\tau_1(.)$ and hence the more mass is assigned. The entire mass distributed on $\Theta\times\{b_1\}$ is equal to the expectation of $\tau_1(.)$; in our case this is $[\tau_1(A)+\tau_1(B)+\tau_1(C)].(1/3)$. This mass distribution induces a (generalized) barycenter, denoted η_1.

Obviously, the entire mass distributed onto $\Theta\times\{b_0\}$ and $\Theta\times\{b_1\}$ is equal to 1. The fact that (η_0,b_0) and (η_1,b_1) are the desired atoms of the extremal upper distribution becomes evident from applying formula (14.68.1) to the above.

The geometric interpretation for the lower extremal distribution works analogously and is illustrated in Fig. 15.2.b .

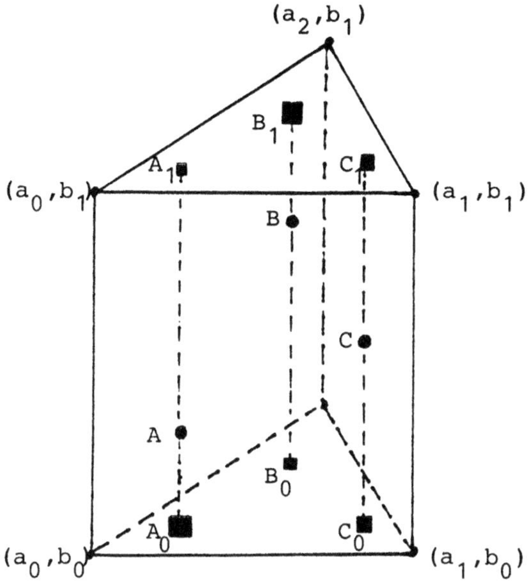

Fig. 15.2.a

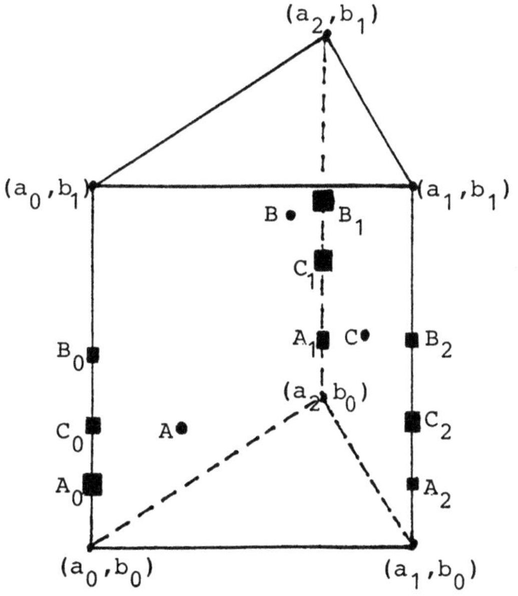

Fig. 15.2.b

Due to the approximation technique presented in the preceding section and illustrated in the above it is evident that these approximations are based, primally and dually, on (generalized) barycenters of simplices with respect to some P-dependent mass distribution . This fact motivates to call our approach *Barycentric Approximation* .

16 *Iterated Approximation and* $\times$*-Simplicial Refinement*

In this section we extend the results of section 14 to the case when stochastic independence is encountered. We suppose to have the probability space given as $(\times_i(\Theta^i\times\Xi^i),\times_i \mathcal{B}^{K_i+L_i},\times_i P^i)$ with Θ^i,Ξ^i **(i=1,...,I) being simplices.** Herewith (trivially), the $\times$-simplex $\Theta\times\Xi=\times_i(\Theta^i\times\Xi^i)$ is compatible with the stochastic independence of the underlying probability space. We set $(\eta,\xi)=\times_i(\eta^i,\xi^i)$, adopt X, $\mathcal{F}(x)$, $\mathcal{F}^c(x)$ as defined in section 14, and consider the expectation functional of the form

$$(E\rho)(x) = \int_{\times_i(\Theta^i\times\Xi^i)} \rho(\times_i(\eta^i,\xi^i),x)\, d(\times_i P^i(\eta^i,\xi^i)) \quad . \tag{16.1}$$

As $\eta^i\in \mathbf{R}^{K_i}$, $\xi^i\in \mathbf{R}^{L_i}$ we additionally suppose that Θ^i, Ξ^i (i=1,...,I) are closed, regular simplices with respect to $\mathbf{R}^{K_i}$, $\mathbf{R}^{L_i}$.

Applying Lemma 14.4 (14.4') and Theorem 14.3 (14.3') with respect to $\Theta^i\times\Xi^i$ and P^i yields the extremal measures $P_{l,i}$, $P_{u,i}$ being given for i=1,...,I with

$$P_{l,i}\,[(a_{\nu_i}, \xi^i_{\nu_i})] = p_{\Theta^i}(\nu_i) \qquad \nu_i=0,\dots,K_i \tag{16.2.1}$$
$$P_{u,i}\,[(\eta^i_{\mu_i}, b_{\mu_i})] = p_{\Xi^i}(\mu_i) \qquad \mu_i=0,\dots,L_i \;. \tag{16.2.2}$$

Here a_{ν_i}, b_{μ_i} denote the (K_i+1), (L_i+1) vertices of Θ^i, Ξ^i. The cross moments of (η^i,ξ^i) defined as in (14.52) with respect to Θ^i, Ξ^i. P^i define the probabilities $p_{\Theta^i}(\nu_i)$, $p_{\Xi^i}(\mu_i)$ according to formulae (14.53.1), (14.65.1) and $\eta^i_{\mu_i}$, $\xi^i_{\nu_i}$ are given through formulae (14.59),(14.66), all with respect to $\Theta^i\times\Xi^i$ and P^i. The counterpart of Δ defined in (14.72) is denoted Δ_i; herewith Δ_i is the set of probability measures coinciding in the cross moments of (η^i,ξ^i), again with respect to $\Theta^i\times\Xi^i$ and P^i .

We state the following

Theorem 16.1 (*Iterated Approximation*): For x in X let $\rho(.,.,x)$ be element of $\mathcal{F}^c(x)$ and the probability space given as $(\times_i(\Theta^i \times \Xi^i),\ \times_i \mathcal{B}^{K_i+L_i},\ \times_i P^i)$ with Θ^i, Ξ^i (i=1,...I) being regular closed simplices (w.r.t. $\mathbf{R}^{K_i}$, $\mathbf{R}^{L_i}$). We denote $P_{l,i}$, $P_{u,i}$ the lower, upper extremal discrete probability measures according to (16.2.1), (16.2.2) and define $P_l = \times_i P_{l,i}$, $P_u = \times_i P_{u,i}$. Then :

$$\int_{\times_i(\Theta^i \times \Xi^i)} \rho(\eta,\xi,x)dP_l(\eta,\xi) \leq \int_{\times_i(\Theta^i \times \Xi^i)} \rho(\eta,\xi,x)dP(\eta,\xi) \leq \int_{\times_i(\Theta^i \times \Xi^i)} \rho(\eta,\xi,x)dP_u(\eta,\xi) \qquad (16.3)$$

Proof : It suffices to outline the proof for I=2. Due to the fact that $P_{l,i} \in \inf^{(s)} \Delta_i$ and due to Fubini's Theorem (see e.g. Halmos 1974 [55, p.148]) we get for the lower approximation :

$$\int_{\times_i(\Theta^i \times \Xi^i)} \rho(\eta,\xi,x)dP = \int_{\times_i(\Theta^i \times \Xi^i)} \rho(\eta,\xi,x)dP^1\, dP^2 \geq \qquad (16.4.1)$$

$$\geq [\ \Sigma_{v_1} \rho(a_{v_1}, \eta^2, \xi^1{}_{v_1}, \xi^2, x).p_{\Theta^1}(v_1)]\ dP^2 \geq \qquad (16.4.2)$$

$$\geq \Sigma_{v_2} [\ \Sigma_{v_1} \rho(a_{v_1}, a_{v_2}, \xi^1{}_{v_1}, \xi^2{}_{v_2}, x).p_{\Theta^1}(v_1)]\ p_{\Theta^2}(v_2) = \qquad (16.4.3)$$

$$= \int_{\Theta^2 \times \Xi^2} \int_{\Theta^1 \times \Xi^1} \rho(\eta,\xi,x)dP_{1,1}\, dP_{1,2} = \int_{\times_i(\Theta^i \times \Xi^i)} \rho(\eta,\xi,x)dP_l \qquad (16.4.4)$$

Upper approximation follows analogously with observing that $P_{u,i} \in \sup^{(s)}\Delta_i$. //

We define $\Delta = \times_i \Delta_i$; i.e. Δ represents the set of those probability measures Q with $Q = \times_i Q^i$ where $Q^i \in \Delta_i$. It is an obvious fact that $P_l \in \inf^{(s)} \Delta$ and $P_u \in \sup^{(s)} \Delta$. The associated dual, the semiinfinite program, maximizes (minimizes) the expectation of lower (upper) approximates for $\rho(.,.,x)$ with respect to all functions being bilinear in (η^i,ξ^i), i=1,...,I , separately.

Remark 16.1: In a similar way, iterated approximation is applicable also to the case where Θ^i,Ξ^i are rectangles. One has simply to use formulae (14.36), (14.19.1), (14.19.2), (14.25.1), (14.25.2) instead of (14.52), (14.53.1), (14.65.1), (14.59), (14.66). In this case we further stress that $P_{l,i}$, $P_{u,i}$ belong to Δ_{Ξ^i}, Δ_{Θ^i} which differ according to (14.36.1) and (14.36.2). As a consequence $P_l \in \times_i \Delta_{\Theta^i} =:$

Δ_Θ and analogously $P_u \in \times_i \Delta_{\Xi^i} =: \Delta_\Xi$. Thus, in the case of working with rectangles we note that, besides the exponential increase in the number of cross moments, the extremal lower and upper probability measures refer to different cross moment conditions, and hence to different feasibility regions.

Exploiting stochastic independences does not impact the computational effort for determining the extremal probability measures P_l and P_u. This is due to the fact that we may evaluate $P_{l,i}$ and $P_{u,i}$ separately for i=1,...,I, requiring totally $\Sigma_i(K_i+1)(L_i+1)$ cross moments. However, stochastic independence impacts of course the effort for integration in (16.4) considerably: there we have to evaluate $\rho(.,.,x)$ at totally $[\Pi_i(K_i+1)+\Pi_i(L_i+1)]$ support points (atoms). This effort is now polynomial in the dimensions of (η^i,ξ^i) with order I.

The idea of improving the approximation of linear stochastic programs with refining the σ-algebra goes back to Kall 1974 [65] and 1980 [68]. Extremal measures given through the classical inequalities of Jensen and Edmundson-Madansky - have been refined via splitting intervals for the convex, stochastic independent case in Huang, Ziemba and Ben-Tal 1977 [62], and in Kall and Stoyan 1982 [79]. From then this strategy has been adopted several times within stochastic programming. We extend that idea with respect to the expectation functional in (16.1) and with respect to the results achieved in section 14 and in the above.

Due to the underlying generality, we are better off with introducing the following notions :

Definition 16.1 : Let Γ denote a simplex. The finite set $D(\Gamma)=\{\Gamma^r \mid \Gamma^r$ is a simplex with $\Gamma^r \subset \Gamma$, r=1,...,R$\}$ is called a **simplicial partition** of Γ, iff

$$\Gamma^{r_1} \cap \Gamma^{r_2} = \emptyset \quad \text{for } r_1 \neq r_2 , \tag{16.5.1}$$
$$\cup_r \Gamma^r = \Gamma . \tag{16.5.2}$$

An element of $D(\Gamma)$ let be called a **subsimplex** of Γ.

Definition 16.2: Let Γ denote a simplex. $\{C^J(\Gamma)$, J=0,1,2,...$\}$ is called a sequence of **nested simplicial collections** of Γ , iff:

$$C^0(\Gamma) := \{\Gamma\} , C^J(\Gamma) := \{ \Gamma^{r,J} \mid \Gamma^{r,J} \subset \Gamma , r=1,...,r^J \} \quad \text{for } J=1,2,... \tag{16.6.1}$$
$$\text{and } \exists\, r(J) \in \{1,...,r^{J-1}\} : \ D(\Gamma^{r(J),J-1}) \subset C^J(\Gamma) \setminus \{\Gamma^{r(J),J-1}\} \tag{16.6.2}$$

where $D(\Gamma^{r(J),J-1})$ represents a simplicial partition of $\Gamma^{r(J),J-1}$.

Definition 16.3: $\mathcal{S}^J$ is called a **×-simplicial partition** of $\Theta\times\Xi$, iff:

$$\mathcal{S}^J := \{ (\Theta\times\Xi)^{t,J} \mid (\Theta\times\Xi)^{t,J} = \times_i (\Theta^{r(i,t,J)}\times\Xi^{s(i,t,J)}) , t=1,...,t^J \text{ and } (\Theta\times\Xi)^{t_1,J}\cap(\Theta\times\Xi)^{t_2,J} = \emptyset \text{ for } t_1 \neq t_2 ; \cup_t(\Theta\times\Xi)^{t,J} = \Theta\times\Xi \} . \tag{16.7.1}$$

An element of $\mathcal{S}^J$ let be called a **sub-×-simplex** of $\Theta\times\Xi$.

Definition 16.4: $\{\mathcal{S}^J , J=0,1,2,...\}$ is called a **×-simplicial refinement scheme** of $\Theta\times\Xi$, iff:

$$\mathcal{S}^J := \{ (\Theta\times\Xi)^{t,J} \mid (\Theta\times\Xi)^{t,J} = \times_i (\Theta^{r(i,t,J)}\times\Xi^{s(i,t,J)}) , t=1,...,t^J \text{ and } (\Theta\times\Xi)^{t_1,J}\cap(\Theta\times\Xi)^{t_2,J} = \emptyset \text{ for } t_1 \neq t_2 ; \cup_t(\Theta\times\Xi)^{t,J} = \Theta\times\Xi \; \Theta^{r(i,t,J)} \in C^J(\Theta^i), \; \Xi^{s(i,t,J)}\in C^J(\Xi^i)\} \tag{16.7.2}$$

for J=0,1,2,..., where $\{C^J(\Theta^i); J=0,1,2,...\}$, $\{C^J(\Xi^i); J=0,1,2,...\}$ are supposed to be a sequence of nested simplicial collections of Θ^i, Ξ^i.

Remark 16.2 : *Simplicial collections* may consist of various *simplicial partitions.* This becomes necessary due to the fact that different subsimplices of Θ^i may be paired with different simplicial partitions of Ξ^i (and vice versa) to form a sub-×-simplex.

Let $P=\times_i P^i$ and J be a fixed iteration. We denote with $P^{i,t,J}$ the conditional probability measure that arises from restricting P^i to $\Theta^{r(i,t,J)}\times\Xi^{s(i,t,J)}$ with $\Theta^{r(i,t,J)}\in C^J(\Theta^i)$, $\Xi^{s(i,t,J)}\in C^J(\Xi^i)$, i=1,2...I, and $P[\times_i(\Theta^{r(i,t,J)}\times\Xi^{s(i,t,J)})]$ supposed to be nonzero. Further let $P_{l,i,t,J}$, $P_{u,i,t,J}$ denote the associated extremal probability measures, that arise if Lemma 14.4 (14.4') and Theorem 14.3 (14.3') are applied w.r.t. $P^{i,t,J}$ and $cl(\Theta^{r(i,t,J)}\times\Xi^{s(i,t,J)})$; this yields supports $\cup_{\nu_i}\{(a_{\nu_i}^{r(i,t,J)}, \xi_{\nu_i}^{s(i,t,J)})\}$, $\cup_{\mu_i}\{(\eta_{\mu_i}^{r(i,t,J)},b_{\mu_i}^{s(i,t,J)})\}$ and probabilities $p_\Theta^{(i,t,J)}(\nu_i)$, $p_\Xi^{(i,t,J)}(\mu_i)$, $\nu_i=0,...,K_i$, $\mu_i=0,...,L_i$, for i=1,...,I according to :

$$P_{l,i,t,J} [(a_{\nu_i}^{r(i,t,J)}, \xi_{\nu_i}^{s(i,t,J)})] = p_\Theta^{(i,t,J)}(\nu_i) \quad i=1,...,I ; t=1,...,t^J \tag{16.8.1}$$
$$P_{u,i,t,J} (\eta_{\mu_i}^{r(i,t,J)},b_{\mu_i}^{s(i,t,J)})] = p_\Xi^{(i,t,J)}(\mu_i) \quad i=1,...,I ; t=1,...,t^J . \tag{16.8.2}$$

Setting $P_{l,t,J}=\times_i P_{l,i,t,J}$, $P_{u,t,J}=\times_i P_{u,i,t,J}$ we obtain discrete probability measures on $(cl[\times_i(\Theta^{r(i,t,J)}\times\Xi^{s(i,t,J)})], \times_i \mathcal{B}^{K_i+L_i})$ through :

$$P_{l,t,J}\ [\times_i(a_{v_i}^{\ r(i,t,J)},\ \xi_{v_i}^{\ s(i,t,J)})] := \Pi_i\ p_\Theta^{(i,t,J)}(v_i)\quad t=1,\ldots,t^J\ , \tag{16.9.1}$$

$$P_{u,t,J}\ [\times_i(\eta_{\mu_i}^{\ r(i,t,J)}, b_{\mu_i}^{\ s(i,t,J)})] := \Pi_i\ p_\Xi^{(i,t,J)}(\mu_i)\quad t=1,\ldots,t^J\ . \tag{16.9.2}$$

Finally, defining

$$P_{l,J} = \Sigma_t\ [P_{l,t,J}\ .\ \Pi_i\, P^i(\Theta^{r(i,t,J)}\times\Xi^{s(i,t,J)})] \tag{16.10.1}$$

$$P_{u,J} = \Sigma_t\ [P_{u,t,J}\ .\ \Pi_i\, P^i(\Theta^{r(i,t,J)}\times\Xi^{s(i,t,J)})] \tag{16.10.2}$$

which is to be understood as

$$P_{l,J}\ [\times_i(a_{v_i}^{\ r(i,t,J)},\ \xi_{v_i}^{\ s(i,t,J)})] = \Pi_i\ p_\Theta^{(i,t,J)}(v_i).P^i(\Theta^{r(i,t,J)}\times\Xi^{s(i,t,J)}) \tag{16.11.1}$$

$$P_{u,J}\ [\times_i(\eta_{\mu_i}^{\ r(i,t,J)}, b_{\mu_i}^{\ s(i,t,J)})] = \Pi_i\ p_\Xi^{(i,t,J)}(\mu_i).P^i(\Theta^{r(i,t,J)}\times\Xi^{s(i,t,J)})\ . \tag{16.11.2}$$

The above yields for J=0,1,..., the desired sequence of **lower** and **upper extremal (discrete) probability measures** on $\times_i\ (\Theta^i\times\Xi^i)$ with respect to the $\times$-simplicial refinement scheme $\{\mathcal{S}^J\ ;\ J=0,1,2,\ldots\}$.

Theorem 16.2 (*Monotonicity of Expectation Functionals*): Let $\{\mathcal{S}^J,\ J=0,1,2,\ldots\}$ denote a $\times$-simplicial refinement scheme of $(\Theta\times\Xi)$ and let $P_{l,J}$, $P_{u,J}$ be defined according to (16.11.1)-(16.11.2). For any x in X let assume $\rho(.,.,x)\in\mathcal{F}^c(x)$. Denoting $(E_{l,J}\ \rho)(x)$, $(E\rho)(x)$, $(E_{u,J}\ \rho)(x)$ the expectation functionals of $\rho(.,.,x)$ with respect to $P_{l,J}$, P, $P_{u,J}$ then it holds on X for J=0,1,2,... :

$$(E_{l,J}\,\rho)(x) \le (E\,\rho)(x) \le (E_{u,J}\,\rho)(x)\ , \tag{16.12}$$

with $\{(E_{l,J}\ \rho)(x),\ J=0,1,2,\ldots\}$ being monotone increasing and $\{(E_{u,J}\ \rho)(x),\ J=0,1,2,\ldots\}$ monotone decreasing.

Proof: We argue dually: $\mathcal{L}^{t,J}(x)$ denotes the set of lower approximates for $\rho(\eta,\xi,x)$ with respect to $(\Theta\times\Xi)^{t,J}$ in the sense of (14.46); analogously $\mathcal{L}^{t',J+1}(x)$, $\mathcal{L}^{t'',J+1}(x)$ is defined w.r.t. $(\Theta\times\Xi)^{t',J+1}$, $(\Theta\times\Xi)^{t'',J+1}$. With $(\Theta\times\Xi)^{t,J}=(\Theta\times\Xi)^{t',J+1}\cup(\Theta\times\Xi)^{t'',J+1}$ then we obviously have

$$\mathcal{L}^{t,J}(x) \subset \mathcal{L}^{t',J+1}(x)\ ,\qquad \mathcal{L}^{t,J}(x) \subset \mathcal{L}^{t'',J+1}(x)\ . \tag{16.13}$$

Hence, maximization of the expectation with respect to $\mathcal{L}^{t',J+1}(x)$, $\mathcal{L}^{t'',J+1}(x)$ (separately) is at least as good as maximization with respect to $\mathcal{L}^{t,J}(x)$, yielding monotonicity of lower expectation functionals due to refinements.
With using the set of upper approximates $\mathcal{U}^{t,J}(x)$, $\mathcal{U}^{t',J+1}(x)$, $\mathcal{U}^{t'',J+1}(x)$ an analogous argument holds for $\{(E_{u,J}\,\rho)(x),\ J=0,1,2,...\}$. //

Remark 16.3 : In the case that $x \rightarrow \rho(\eta,\xi,x)$ is convex on X for any (η,ξ) in $\Theta\times\Xi$ we obtain with the approximate expectation functionals $(E_{l,J}\,\rho)(x)$ and $(E_{u,J}\,\rho)(x)$ *outer* and *inner approximations* for $(E\,\rho)(x)$ in every iteration J.

To illustrate how to generate a $\times$-simplicial refinement scheme we outline one iteration in detail: Suppose we have with $\mathcal{S}^J$ a $\times$-simplicial partition of $\Theta\times\Xi$; (in the case J=0 we set $\mathcal{S}^0=\{\Theta\times\Xi\}$). To ease the exposition we suppose to have simplices with Θ^i, Ξ^i; we fix t and choose the sub-$\times$-simplex $(\Theta\times\Xi)^{t,J}\in\mathcal{S}^J$ for splitting. Having $(\Theta\times\Xi)^{t,J} = \times_i\ (\Theta^{r(i,t,J)}\times\Xi^{s(i,t,J)})$ yields :

#1 :
Determine the pair of vertices among the simplices $\{\Theta^{r(i,t,J)},\ \Xi^{s(i,t,J)}\}$ with the longest edge; we take the line-segment $\{a_{v'_{i'}}^{\ r(i',t,J)},\ a_{v''_{i'}}^{\ r(i',t,J)}\}$ of $\Theta^{r(i',t,J)}$ as longest edge - without loss of generality . (16.14)

#2 :
Split the edge $[a_{v'_{i'}}^{\ r(i',t,J)},\ a_{v''_{i'}}^{\ r(i',t,J)}\ ]$ in the middle and denote the corresponding point with $a_{v'''_{i'}}$; denoting the hyperplane induced by the $K_{i'}$ (affinely independent) points

$$\{a_{v'''_{i'}},\ a_{v_{i'}}^{\ r(i',t,J)}\ (v_{i'}=0,...,K_{i'})\}\backslash\{a_{v'_{i'}}^{\ r(i',t,J)},a_{v''_{i'}}^{\ r(i',t,J)}\} \tag{16.15.1}$$

with $H(\eta^{i'})=h^{i'}$ and setting $|C^J(\Theta^{i'})|=:r^{i',J}$ for the cardinality of $C^J(\Theta^{i'})$, then the new subsimplices within $C^{J+1}(\Theta^{i'})$ are given according to

$$\Theta^{i',r'} := \Theta^{r(i',t,J)} \cap \{\eta^{i'}\ |\ H(\eta^{i'}) \leq h^{i'}\ \} \tag{16.15.2}$$

$$\Theta^{i',r''} := \Theta^{r(i',t,J)} \cap \{\eta^{i'}\ |\ H(\eta^{i'}) > h^{i'}\ \} \tag{16.15.3}$$

and

$$C^{J+1}(\Theta^{i'}) = C^J(\Theta^{i'})\cup\{\Theta^{i',r'}\ ,\Theta^{i',r''}\} \tag{16.15.4}$$

with $r'=r^{i',J}+1$ and $r''=r^{i',J}+2$.

#3 :
Remove $(\Theta\times\Xi)^{t,J}$ from $\mathcal{S}^J$, i.e. $\mathcal{S}^{J+1} = \mathcal{S}^J \setminus (\Theta\times\Xi)^{t,J}$, and

$$\mathcal{S}^{J+1} := \mathcal{S}^{J+1} \cup \{\times_i \, (\Theta^{r'(i,t,J+1)} \times \Xi^{s(i,t,J)}) \, , \times_i \, (\Theta^{r''(i,t,J)} \times \Xi^{s(i,t,J)})\} \tag{16.16.1}$$

where

$$\text{for } i \neq i' : \Theta^{r'(i,t,J+1)} = \Theta^{r(i,t,J)} \quad , \tag{16.16.2}$$

$$\text{for } i = i' : \Theta^{r'(i,t,J+1)} = \Theta^{i',r'} \, , \, \Theta^{r''(i,t,J+1)} = \Theta^{i',r''} \, . \tag{16.16.3}$$

The above steps #1-#3 may of course be applied to all sub-$\times$-simplices of $\mathcal{S}^J$, and not just to one. It is easy to verify that the so obtained sequence $\{\mathcal{S}^J, J=0,1,2,...\}$ satisfies the conditions in (16.7). Finally we want to stress that through the characterization of the ('half-open') subsimplices according to (16.15.1), (16.15.2) all the sub-$\times$-simplices - generated within $\{\mathcal{S}^J, J=0,1,2,...\}$ - remain measurable.

Due to the nature of approximations derived with respect to a $\times$-simplicial refinement scheme, we shall call the associated *sequence of extremal bilinear functions*, the associated *sequence of extremal probability measures* respectively, **barycentric approximation scheme**.

Remark 16.3 : If Θ^i, Ξ^i are rectangles, we simply treat the corresponding intervals as one-dimensional simplices in Definitions 16.2, 16.3 correspondingly; applying #1-#3 yields a *rectangular refinement*. The hyperplane in (16.17) then simplifies to $\eta_k^{i'} = h^{i'}$ where $\eta_k^{i'}$ is the k-th component of $\eta^{i'}$. However, the price we pay for these simplifications is an exponential increase in cross moment evaluations.

Remark 16.4: We have to mention, that simplicial partitions have achieved considerable attention e.g. in global optimization. Subdivision strategies are investigated in Horst and Tuy 1990 [61], and Tuy 1991 [139], based on socalled *radial subdivision* of a simplex; therein it is allowed to take any point within a simplex for partitioning, obtaining herewith (in general) more than 2 subsimplices.

In this section we specify and discuss the scope within we apply the barycentric approximation to stochastic two-stage programs.

Let the probability space be given as $(\Theta\times\Xi,\mathcal{B}^{K+L},P)=(\times_i(\Theta^i\times\Xi^i),\times_i \mathcal{B}^{K_i+L_i},\times_i P^i)$. Due to the results available from sections 4,5,6,14 and 16 we consider stochastic two-stage problems of the form and in accordance with the notation used therein :

$$\inf_{x\in X} \quad f(x) + \int_{\times_i(\Theta^i\times\Xi^i)} \rho(\times_i(\eta^i,\xi^i),x)\, d(\times_i P^i(\eta^i,\xi^i)) , \tag{17.1.1}$$

or equivalently with $(\eta,\xi)=\times_i(\eta^i,\xi^i)$,

$$\inf_{x\in X} \quad f(x) + \int_{\Theta\times\Xi} \rho(\eta,\xi,x)\, dP(\eta,\xi) \tag{17.1.2}$$

where the following conditions are supposed to be satisfied :

(C1) $X \subset \mathbf{R}^n$ closed and non-empty; f: $\mathbf{R}^n\to\mathbf{R}$ lower semicontinuous proper ;
(C2) $\Theta\times\Xi := \times_i (\Theta^i\times\Xi^i)$, where $\Theta\subset\mathbf{R}^K$, $\Xi\subset\mathbf{R}^L$, $\Theta^i\subset\mathbf{R}^{K_i}$, $\Xi^i\subset\mathbf{R}^{L_i}$ and Θ^i , Ξ^i representing regular and closed simplices for i=1,...,I ;
(C3) $\rho(\eta,\xi,.)$: $\mathbf{R}^n\to\mathbf{R}\cup\{+\infty\}$ lower semicontinuous, finite on X and $+\infty$ outside X (hence proper) , for every $(\eta,\xi)\in\Theta\times\Xi$;
(C4) $\rho(.,.,x)$: $\Theta\times\Xi\to\mathbf{R}$ is a subdifferentiable (concave-convex) saddle function for all $x\in X$;
(C5) $E\rho$: $\mathbf{R}^n\to\mathbf{R}\cup\{+\infty\}$ proper, lower semicontinuous .

In view of *uniform integrability* in the stability result (Proposition 3.1), condition (C2) is not too restrictive: in Proposition 3.1 the events outside a compact set are supposed to be negligible with respect to a set of probability measures and with respect to the collection $\{\rho(.,.,x)|\ x\in X\}$; in (C2) we additionally assume to have a $\times$-simplex, compatible with the stochastic independent subvectors (η^i,ξ^i), i=1,...,I.

In the case that $\rho(\eta,\xi,.)$: $\mathbf{R}^n\to\mathbf{R}\cup\{+\infty\}$ are convex for every $(\eta,\xi)\in\Theta\times\Xi$ in (C3), then (C5) holds self-acting (Proposition 1.1). However, in the nonconvex case we have to require additional assumptions for (C5) to hold: lower semicontinuity of the expectation functional $E\rho$ is ensured if either additionally the collections

$\{\rho(.,.,x) \mid x \in \mathbf{R}^n\}$ have an integrable minorant - independent of x (Proposition 2.2 in Wets 1989 [144]), or if the collection $\{\rho(\eta,\xi,.) \mid (\eta,\xi) \in \Theta\times\Xi\}$ is uniformly equicontinuous on X; the latter case implies even continuity of $x \to (E\rho)(x)$ relativ to X (Wets 1989 [144]) .

In the case that $(\eta,\xi) \to \rho(\eta,\xi,x)$ represents the optimal value function of an inner convex optimization problem with (η,ξ) as random data and with x as parameter we consider:

$$\rho(\eta,\xi,x) = \inf_y \phi(\eta,\xi,y,x) \tag{17.2}$$

where $\phi(.,.,.,.)$: $\Theta\times\Xi\times\mathbf{R}^{n'}\times\mathbf{R}^n \to \mathbf{R}$ is supposed to be a normal integrand. In order that $\rho(\eta,\xi,x)$ is lower semicontinuous we require an appropriate *growth condition* (Rockafellar 1976 [121, p.180]) to hold: If for every $(\eta,\xi) \in \Theta\times\Xi$, every $\alpha \in \mathbf{R}$ and every bounded set $K \subset \mathbf{R}^n$ the set $\{y \in \mathbf{R}^{n'} \mid \exists\, x \in K \text{ with } \phi(\eta,\xi,y,x) \leq \alpha\}$ is bounded, then the minimum in (17.2) is attained and $\rho(\eta,\xi,x)$ is lower semicontinuous in x, and hence a normal integrand.

In view of Definition 10.2 we additionally assume $\phi(.,.,.,x)$: $\Theta\times\Xi\times\mathbf{R}^{n'} \to \mathbf{R}$ to be lower closed, concave in η and convex in (ξ,y), for every x in X. This allows to reformulate (C4) as

(C4)' For every x in X , the convex program (17.2) is stable for all $(\eta,\xi) \in \Theta\times\Xi$.

Remark 17.1 : In accordance with the previous sections let denote the vertices of the $\times$-simplex $\Theta\times\Xi$ with (a_ν,b_μ) where $\nu=(\nu_1,\ldots,\nu_I)$, $\mu=(\mu_1,\ldots,\mu_I)$ with $\nu_i=0,\ldots,K_i$ and $\mu_i=0,\ldots,L_i$ for $i=1,\ldots,I$.

Applying Theorem 10.2 (adapting the notation with respect to (17.2)) we state

Lemma 17.1: Let x be arbitrary but fixed in X, with $\phi(.,.,.,x)$: $\Theta\times\Xi\times\mathbf{R}^{n'} \to \mathbf{R}$ being a lower closed saddle function and $\Theta\times\Xi = \times_i (\Theta^i\times\Xi^i)$ denoting a $\times$-simplex. For (17.2) being stable at any $(\eta,\xi) \in (\Theta\times\Xi)$ it is sufficient that for every vertex $(a_\nu,b_\mu) \in (\Theta\times\Xi)$ it holds:

$\exists\, y(a_\nu,b_\mu,x)$ so that $\phi(.,.,y(a_\nu,b_\mu,x),x)$ is continuous and finite at (a_ν,b_μ); (17.3.1)

$\exists\, U_{1,x,\nu,\mu} \in \mathcal{N}(\eta=a_\nu)$: $\rho(\eta,b_\mu,x)$ finite $\forall\, \eta \in U_{1,x,\nu,\mu}$. (17.3.2)

Proof: We outline the proof for I=1 and Θ,Ξ being simplices. We have to show that $\rho(\eta,\xi,x)$ is subdifferentiable at any $(\eta,\xi)\in(\Theta\times\Xi)$ (Note that x is kept fixed but arbitrary in X).

i) $(\eta^0,\xi^0)\in \mathrm{int}(\Theta\times\Xi)$: As stability of (17.2) on (a_v,b_μ) implies subdifferentiability of $\rho(.,.,x)$ at (a_v,b_μ) we construct (bilinear) lower and upper approximates $L(\eta,\xi,x)$ and $U(\eta,\xi,x)$ for $\rho(\eta,\xi,x)$ on $(\Theta\times\Xi)$, as has been done in section 14. Hence $\rho(\eta,\xi,x)$ is finite on $(\Theta\times\Xi)$ and so $(\eta^0,\xi^0)\in \mathrm{int}(\Theta\times\Xi)\subset \mathrm{dom}\rho(.,.,x)$. Therefore $\rho(\eta,\xi,x)$ is subdifferentiable at $(\eta^0,\xi^0)\in \mathrm{int}(\Theta\times\Xi)$ and herewith (17.2) stable at any $(\eta^0,\xi^0)\in \mathrm{int}(\Theta\times\Xi)$.

ii) $(\eta^0,\xi^0)\in \mathrm{bd}(\Theta\times\Xi)$: We shall prove that we may find points a_v^0 and b_μ^0 within a neighborhood of the vertices a_v and b_μ, with $\rho(.,.,x)$ being stable at (a_v^0,b_μ^0) and $(\eta^0,\xi^0)\in \mathrm{int}(\Theta^0\times\Xi^0)$, whereby $\Theta^0\times\Xi^0$ denotes the $\times$-simplex defined by (a_v^0,b_μ^0).

As $\Theta\times\Xi$ is a $\times$-simplex, $(\eta^0,\xi^0)\in \mathrm{bd}(\Theta\times\Xi)$ is equivalent with the existence of some barycentric weights $\lambda_v(\eta^0)=0$ or $\tau_\mu(\xi^0)=0$. Without loss of generality we start with $\lambda_v(\eta^0)=0$ (for some $v\in\{1,...,K\}$) and let $\lambda_0(\eta^0)$ be strictly positive. Defining

$$\eta^B = [\Sigma_v\, a_v\,]/[K+1] \tag{17.4.1}$$

and denoting with $[a_0, \eta^B\,]$ an associated line-segment we may choose some point a_0^0 so that $a_0\in \mathrm{ri}[a_0^0,\eta^B]$. This implies that $a_0\in \mathrm{int}[\mathrm{co}\{a_0^0,a_v(v=1,...,K)\}]$. For convenience let $\Theta^0:=\mathrm{co}\{a_0^0, a_v\ (v=1,...,K)\}$.
Analogously, we obtain for b_0 with strictly positive $\tau_0(\xi^0)$ a distinguished point b_0^0 with $b_0\in \mathrm{ri}[b_0^0, \xi^B\,]$ where

$$\xi^B = [\Sigma_\mu\, b_\mu\,]/[L+1]. \tag{17.4.2}$$

Again, $\Xi^0 := \mathrm{co}\{b_0^0, b_\mu\ (\mu=1,...,L)\}$.

Due to continuity of $\phi(.,.,y(a_0,b_0,x),x)$ at (a_0,b_0) in some open neighborhood and finiteness of $\rho(\eta,b_0,x)$ in an open neighborhood (being subset of $U_{1,x,v,\mu}\in\mathcal{N}(\eta=a_0)$) we choose a_0^0 and b_0^0 sufficiently close to a_0 and b_0 on the associated line-segment, so that (17.3.1) and (17.3.2) still hold for (a_0^0,b_0^0). As all barycentric weights of η^0,ξ^0 (now!) with respect to Θ^0,Ξ^0 are strictly positive we have $(\eta^0,\xi^0)\in \mathrm{int}(\Theta^0\times\Xi^0)$. Due to construction (17.2) is stable at the vertices of $\Theta^0\times\Xi^0$. Applying the same argumentation as in part i) we obtain subdifferentiability of $\rho(.,.,x)$ at (η^0,ξ^0) and hence stability of (17.2) at (η^0,ξ^0).

If I>1 we have to apply the same idea to all simplices Θ^i,Ξ^i (i=1,...,I). Lower and upper approximates are then bilinear in (η^i,ξ^i) separately, enforcing finiteness and continuity of $\rho(.,.,x)$ in an open set covering the entire $\times$-simplex. //

(C3) entails $X\subset \mathrm{dom}_x\ \rho(\eta,\xi,.)$ for every $(\eta,\xi)\in(\Theta\times\Xi)$, we cover the common assumption within stochastic programming, that no induced constraints arise through the second-stage problem. Moreover, subdifferentiability in (C4) or in (C4)' reduces to complete recourse and dual feasibility almost everywhere within linear stochastic two-stage programs, implying (C3) self-acting and (C5) with the existence of cross moments of (η,ξ) - see Wets 1974 [141] and Kall 1976 [66] .

Under the assumptions (C1),(C2),(C3),(C4) - or (C4)' - and (C5) with using the extremal measures of (16.11.1), (16.11.2) for some $\times$-simplicial partition $\mathcal{S}^J$ of $\Theta\times\Xi$, we obtain with

$$\inf_{x\in X}\ f(x) + \int_{\Theta\times\Xi} \rho(\eta,\xi,x)dP_{l,J}(\eta,\xi) \qquad (17.5.1)$$

$$\inf_{x\in X}\ f(x) + \int_{\Theta\times\Xi} \rho(\eta,\xi,x)dP_{u,J}(\eta,\xi) \qquad (17.5.2)$$

approximate problems for (17.1) with satisfying (C1)-(C5).

18 Convergence of Approximations

Below, we state convergence properties of the approximate problems (17.5). In particular, we are interested in whether

$$\lim_{J\to\infty}\{\inf_{x\in X} f(x)+ \int_{\Theta\times\Xi}\rho(\eta,\xi,x)dP_{l,J}\} = \inf_{x\in X}\ f(x) + \int_{\Theta\times\Xi}\rho(\eta,\xi,x)dP(\eta,\xi) \quad (18.1.1)$$

$$\lim_{J\to\infty}\{\inf_{x\in X} f(x)+ \int_{\Theta\times\Xi}\rho(\eta,\xi,x)dP_{u,J}\} = \inf_{x\in X}\ f(x) + \int_{\Theta\times\Xi}\rho(\eta,\xi,x)dP(\eta,\xi) \quad (18.1.2)$$

and whether the approximate solutions accumulate to some solution of (17.1).

For convenience we use for J=0,1,2,...

$$\psi_{l,J}(x) := f(x) + \int_{\Theta\times\Xi} \rho(\eta,\xi,x)dP_{l,J}(\eta,\xi) = f(x) + (E_{l,J}\,\rho)(x) \tag{18.2.1}$$

$$\psi_{u,J}(x) := f(x) + \int_{\Theta\times\Xi} \rho(\eta,\xi,x)dP_{u,J}(\eta,\xi) = f(x) + (E_{u,J}\,\rho)(x) \tag{18.2.2}$$

$$\psi(x) := f(x) + \int_{\Theta\times\Xi} \rho(\eta,\xi,x)dP(\eta,\xi) = f(x) + (E\,\rho)(x) \tag{18.2.3}$$

$$\Psi_{l,J} := \operatorname{argmin}_{x\in X}\ \psi_{l,J}(x) \tag{18.2.4}$$

$$\Psi_{u,J} := \operatorname{argmin}_{x\in X}\ \psi_{u,J}(x) \tag{18.2.5}$$

$$\Psi := \operatorname{argmin}_{x\in X}\ \psi(x)\ . \tag{18.2.6}$$

We state

Theorem 18.1: Let $\{\mathcal{S}^J, J=0,1,2,...\}$ represent a $\times$-simplicial refinement scheme with the property that the sub-$\times$-simplices become arbitrarily small (with respect to its diameter) and suppose that (C1)-(C5) hold. Let $P_{l,J}$ and $P_{u,J}$ denote the extremal measures associated with $\{\mathcal{S}^J, (J=0,1,2,...)\}$ according to (16.11.1) and (16.11.2). Then $\{\psi_{l,J}(x), J=0,1,2,...\}$, $\{\psi_{u,J}(x), J=0,1,2,...\}$ epi-converge to $\psi(x)$.

Proof: As the diameters of all sub-$\times$-simplices $(\Theta\times\Xi)^{\iota,J} \in \mathcal{S}^J$ $(J=0,1,2,...)$ tend to 0 with $J\rightarrow\infty$, and due to the construction of $P_{l,J}$, $P_{u,J}$ according to (16.10.1), (16.10.2) we have weak convergence of $\{P_{l,J}, J=0,1,2,...\}$ and $\{P_{u,J}, J=0,1,2,...\}$ to P. (C4) and compactness of $\Theta\times\Xi$ yield boundedness and continuity of $\rho(.,.,x)$ for any x in X. These facts and (C3) imply that both $\{(E_{l,J}\rho)(x), J=0,1,2,...\}$ and $\{(E_{u,J}\rho)(x), J=0,1,2,...\}$ converge pointwise to $(E\rho)(x)$. Monotonicity of $\{(E_{l,J}\rho)(x), J=0,1,2,...\}$ and $\{(E_{u,J}\,\rho)(x), J=0,1,2,...\}$ - due to Theorem 16.2 - together with lower semicontinuity of f(x) and $(E\rho)(x)$ yields epi-convergence of $\{\psi_{l,J}(x), J=0,1,2...\}$ and $\{\psi_{u,J}(x), J=0,1,2,...\}$ (Proposition 4.3). //

We want to stress that in contrast to Proposition 4.3 there is no need of local Lipschitz continuity of $\rho(\eta,\xi,x)$ in x. This is due to the fact that $\{P_{l,J}, J=0,1,2,...\}$ and $\{P_{u,J}, J=0,1,2,...\}$ are weak convergent, extremal measures entailing monotonicity of $\{(E_{l,J}\rho)(x), J=0,1,2,...\}$ and of $\{(E_{u,J}\,\rho)(x), J=0,1,2,...\}$.

With Theorem 18.1 and Proposition 4.2 we have the assurance that any accumulation point of the approximate solutions $\{\Psi_{l,J}, J=0,1,2...\}$ and $\{\Psi_{u,J}, J=0,1,2,...\}$ solve the original problem (17.1), and in that case

$$\liminf_{J\rightarrow\infty} \psi_{l,J} = \liminf_{J\rightarrow\infty} \psi_{u,J} = \inf \psi\ . \tag{18.3}$$

Due to the results of section 14 and taking Fig. 15.1. as illustration, we know that the barycentric approximation is the better, the 'more linear' the function $(\eta,\xi)\rightarrow\rho(\eta,\xi,x)$ is for some iterate x; in particular we obtain exactness of the approximation with respect to some sub-×-simplex, if $(\eta,\xi)\rightarrow\rho(\eta,\xi,x)$ is linear on that sub-×-simplex. This motivates to discover, how 'close' $\rho(\eta,\xi,x)$ is to a linear function.

With using ×-simplicial refinements according to (16.14)-(16.16) and having some sub-×-simplex $(\Theta\times\Xi)^{t,J}$ to split, we are free in choosing the simplex and one of its edge for splitting. This fact gives rise to focus on linearity/nonlinearity of $\rho(\eta,\xi,x)$ with respect to **all** edges of the sub-×-simplex $(\Theta\times\Xi)^{t,J}$ that is to split: partitioning that edge which is 'most nonlinear' (with respect to all edges of $(\Theta\times\Xi)^{t,J}$), we may expect a significant improvement. We perform such an analysis via evaluating the duality gaps through the subgradients of $\rho(\eta,\xi,x)$ with respect to (η,ξ). In the case that $\rho(\eta,\xi,x)$ is the optimal value function of some inner optimization problem this amounts to use the dual solutions, supposed to exist according to (C4)'. In fact, that which follows may be seen as a generalization of the refinement strategy proposed in Frauendorfer and Kall 1988 [49] and applied for purely convex stochastic linear programs with stochastically independent random components. Extensive tests therein empirically illustrate the importance of employing duality analysis for the refinement. Due to this fact we extend the corresponding strategy to the saddle case.

To simplify the exposition we restrict on the case I=1 and on Θ, Ξ being simplices, keep J fixed but arbitrary and suppose to have with x^J a minimizer of the lower approximating problem (17.5.1); we therefore may omit 'i' as superscript in the notations used in section 16; the counterpart for I>1 is purely technical and provides no more insight.

For any cell $(\Theta\times\Xi)^{t,J}\in\mathcal{S}^J$ we linearize $\rho(.,.,x^J)$ due to its subdifferentiability - ensured by (C4) or (C4)' - at $(a_v^{r(t,J)},\xi_v^{s(t,J)})$, $(\eta_\mu^{r(t,J)},b_\mu^{s(t,J)})$ according to :

$$\begin{aligned}(\eta,\xi)\rightarrow\rho(a_v^{r(t,J)},\xi_v^{s(t,J)},x^J)+&\\ +<\eta^*(a_v^{r(t,J)},\xi_v^{s(t,J)},x^J),\eta-a_v^{r(t,J)}>+&\\ +<\xi^*(a_v^{r(t,J)},\xi_v^{s(t,J)},x^J),\xi-\xi_v^{s(t,J)}>&\end{aligned}\qquad(19.1.1)$$

$$(\eta,\xi) \rightarrow \rho(\eta_\mu^{r(t,J)}, b_\mu^{s(t,J)}, x^J) + \\ + <\eta^*(\eta_\mu^{r(t,J)}, b_\mu^{s(t,J)}, x^J), \eta - \eta_\mu^{r(t,J)}> + \\ + <\xi^*(\eta_\mu^{r(t,J)}, b_\mu^{s(t,J)}, x^J), \xi - b_\mu^{s(t,J)}> \qquad (19.1.2)$$

with $(\eta^*(a_\nu^{r(t,J)},\xi_\nu^{s(t,J)},x^J),\xi^*(a_\nu^{r(t,J)},\xi_\nu^{s(t,J)},x^J))$, $(\eta^*(\eta_\mu^{r(t,J)},b_\mu^{s(t,J)},x^J),\xi^*(\eta_\mu^{r(t,J)},b_\mu^{s(t,J)},x^J))$ representing the subgradients of $\rho(.,.,x^J)$ at $(a_\nu^{r(t,J)},\xi_\nu^{s(t,J)})$ and $(\eta_\mu^{r(t,J)}, b_\mu^{s(t,J)})$. If we have in particular either

$$\rho(a_{\nu'}^{r(t,J)}, \xi_{\nu'}^{s(t,J)}, x^J) \cong \rho(a_\nu^{r(t,J)}, \xi_\nu^{s(t,J)}, x^J) + \\ + <\eta^*(a_\nu^{r(t,J)},\xi_\nu^{s(t,J)},x^J), a_{\nu'}^{r(t,J)} - a_\nu^{r(t,J)}> + \\ + <\xi^*(a_\nu^{r(t,J)},\xi_\nu^{s(t,J)},x^J), \xi_{\nu'}^{s(t,J)} - \xi_\nu^{s(t,J)}> \qquad (19.2.1)$$

or

$$\rho(a_\nu^{r(t,J)}, \xi_\nu^{s(t,J)}, x^J) \cong \rho(a_{\nu'}^{r(t,J)}, \xi_{\nu'}^{s(t,J)}, x^J) + \\ + <\eta^*(a_{\nu'}^{r(t,J)},\xi_{\nu'}^{s(t,J)},x^J), a_\nu^{r(t,J)} - a_{\nu'}^{r(t,J)}> + \\ + <\xi^*(a_{\nu'}^{r(t,J)},\xi_{\nu'}^{s(t,J)},x^J), \xi_\nu^{s(t,J)} - \xi_{\nu'}^{s(t,J)}> \qquad (19.2.2)$$

this let us expect that $\rho(.,.,x^J)$ is 'almost linear' on the line-segment $[(a_\nu^{r(t,J)},\xi_\nu^{s(t,J)}),(a_{\nu'}^{r(t,J)},\xi_{\nu'}^{s(t,J)})]$; the analogous argument holds of course for $[(\eta_\mu^{r(t,J)},b_\mu^{s(t,J)}),(\eta_{\mu'}^{r(t,J)}, b_{\mu'}^{s(t,J)})]$ with either

$$\rho(\eta_{\mu'}^{r(t,J)}, b_{\mu'}^{s(t,J)}, x^J) \cong \rho(\eta_\mu^{r(t,J)}, b_\mu^{s(t,J)}, x^J) + \\ + <\eta^*(\eta_\mu^{r(t,J)},b_\mu^{s(t,J)},x^J), \eta_{\mu'}^{r(t,J)} - \eta_\mu^{r(t,J)}> + \\ + <\xi^*(\eta_\mu^{r(t,J)},b_\mu^{s(t,J)},x^J), b_{\mu'}^{s(t,J)} - b_\mu^{s(t,J)}> \qquad (19.3.1)$$

or

$$\rho(\eta_\mu^{r(t,J)}, b_\mu^{s(t,J)}, x^J) \cong \rho(\eta_{\mu'}^{r(t,J)}, b_{\mu'}^{s(t,J)}, x^J) + \\ + <\eta^*(\eta_{\mu'}^{r(t,J)},b_{\mu'}^{s(t,J)},x^J), \eta_\mu^{r(t,J)} - \eta_{\mu'}^{r(t,J)}> + \\ + <\xi^*(\eta_{\mu'}^{r(t,J)},b_{\mu'}^{s(t,J)},x^J), b_\mu^{s(t,J)} - b_{\mu'}^{s(t,J)}> . \qquad (19.3.2)$$

So, defining $\Delta(\nu,\nu',x^J)$, $\Delta(\mu,\mu',x^J)$ according to

$$\Delta(\nu,\nu',x^J) := \min\{ |\rho(a_\nu^{r(t,J)},\xi_\nu^{s(t,J)},x^J) + <\eta^*(a_\nu^{r(t,J)},\xi_\nu^{s(t,J)},x^J), a_{\nu'}^{r(t,J)} - a_\nu^{r(t,J)}> \\ + <\xi^*(a_\nu^{r(t,J)},\xi_\nu^{s(t,J)},x^J), \xi_{\nu'}^{s(t,J)} - \xi_\nu^{s(t,J)}> - \rho(a_{\nu'}^{r(t,J)},\xi_{\nu'}^{s(t,J)},x^J)| , \\ |\rho(a_{\nu'}^{r(t,J)}, \xi_{\nu'}^{s(t,J)}, x^J) + <\eta^*(a_{\nu'}^{r(t,J)},\xi_{\nu'}^{s(t,J)},x^J), a_\nu^{r(t,J)} - a_{\nu'}^{r(t,J)}> + \\ + <\xi^*(a_{\nu'}^{r(t,J)},\xi_{\nu'}^{s(t,J)},x^J), \xi_\nu^{s(t,J)} - \xi_{\nu'}^{s(t,J)}> - \rho(a_\nu^{r(t,J)}, \xi_\nu^{s(t,J)}, x^J)| \} \qquad (19.4.1)$$

$$\begin{aligned}\Delta(\mu,\mu',x^J):=\min\{ & |\rho(\eta_\mu^{r(t,J)},b_\mu^{s(t,J)},x^J)+\langle\eta^*(\eta_\mu^{r(t,J)},b_\mu^{s(t,J)},x^J),\eta_{\mu'}^{r(t,J)}-\eta_\mu^{r(t,J)}\rangle+ \\ & +\langle\xi^*(\eta_\mu^{r(t,J)}, b_\mu^{s(t,J)},x^J),b_{\mu'}^{s(t,J)}-b_\mu^{s(t,J)}\rangle-\rho(\eta_{\mu'}^{r(t,J)}, b_{\mu'}^{s(t,J)},x^J)|, \\ & |\rho(\eta_{\mu'}^{r(t,J)},b_{\mu'}^{s(t,J)},x^J)+\langle\eta^*(\eta_{\mu'}^{r(t,J)},b_{\mu'}^{s(t,J)},x^J),\eta_\mu^{r(t,J)}-\eta_{\mu'}^{r(t,J)}\rangle+ \\ & +\langle\xi^*(\eta_{\mu'}^{r(t,J)}, b_{\mu'}^{s(t,J)},x^J),b_\mu^{s(t,J)}-b_{\mu'}^{s(t,J)}\rangle-\rho(\eta_\mu^{r(t,J)},b_\mu^{s(t,J)},x^J)|\}\end{aligned} \qquad (19.4.2)$$

we interpret $\Delta(\nu,\nu',x^J)$, $\Delta(\mu,\mu',x^J)$ as an indicator for the degree of nonlinearity of $\rho(.,.,x^J)$ along the associated line segments.

For example, in Fig. 19.1. we realize that the duality gap with respect to the convex argument (Fig. 19.1.a) is much larger than the duality gap with respect to the concave argument (Fig. 19.1.b). This tells us that we very likely perform better with splitting the edge of the (1-dimensional) simplex associated with the convex argument.

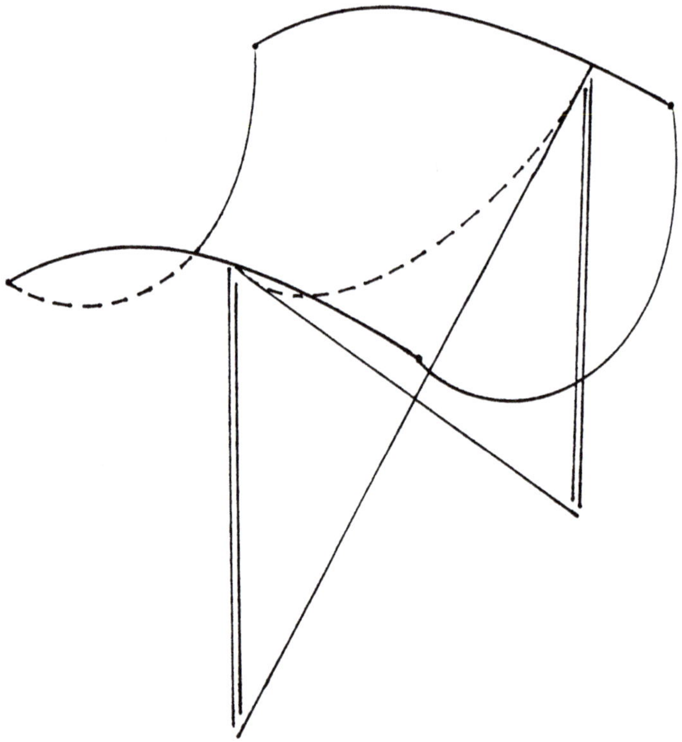

Fig. 19.1.a

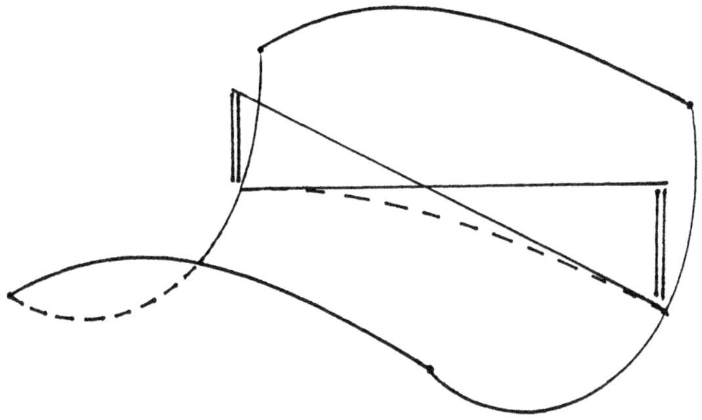

Fig. 19.1.b

It is now at hand that we modify #1 in (16.14) and define :

#1' Split that edge of the $\times$-simplex $(\Theta\times\Xi)^{t,J}$ in the middle, whose adjacent vertices solve

$$\max_{\{(\nu,\nu')(\mu,\mu')\}} \{\Delta(\nu,\nu',x^J) , \Delta(\mu,\mu',x^J)\} \quad . \tag{19.5}$$

Remark 19.1: An edge of a $\times$-simplex connects two adjacent vertices. For simplices, any pair of vertices is adjacent. This fact strongly simplifies the evaluation of (19.5) - see chapter V for more details.

According to the assumptions of Theorem 18.1 we have to ensure that the diameter of the sub-$\times$-simplices tend to 0; as in certain rather peculiar cases, (19.5) does not suffice to force the diameters to become arbitrarily small, we <u>must not</u> ignore #1 in (16.8) completely; in the proposed iterative completion - described below - this necessitates to incorporate #1 in a certain conditional sense: we apply #1 instead of #1', if the proportion of the longest to the shortest edge - subject to the underlying sub-$\times$-simplex going to be split - exceeds a prescribed tolerance.

Based on the notations introduced in the previous sections, assuming (C1)-(C4), or (C4)', and (C5) of section 17 to hold, defining the accuracy level ε posed on the stochastic programming problem, defining a tolerance level ε^0 with respect to which we control the diameters of the splitting sub-$\times$-simplices (according to Remark 19.2), we embed the *barycentric approximation scheme* within an iterative completion, designed to solve stochastic two-stage problems (17.1):

#1 : (20.1.1)

$$J:=0;\ r^{i,J}:=1,\ C^J(\Theta^i)=\{\Theta^i\}=\{\Theta^{r(i,J)} \mid \Theta^{r(i,J)} \subset \Theta^i \text{ for } r(i,J)=1,\dots,r^{i,J}\}, i=1,\dots,I$$
$$s^{i,J}:=1,\ C^J(\Xi^i)=\{\Xi^i\}=\{\Xi^{s(i,J)} \mid \Xi^{s(i,J)} \subset \Xi^i \text{ for } s(i,J)=1,\dots,s^{i,J}\}, i=1,\dots,I$$
$$t^J:=1,\ \mathcal{S}^J:=\{\Theta\times\Xi\}=\{(\Theta\times\Xi)^{t,J} \mid t=1,\dots,t^J\}$$

#2: (20.1.2)

Determine w.r.t. $\mathcal{S}^J=\{(\Theta\times\Xi)^{t,J} \mid t=1,\dots,t^J\}$:

$P_{l,i,t,J}$, $P_{u,i,t,J}$ with respect to $\Theta^{r(i,t,J)}$, $\Xi^{s(i,t,J)}$ defining $(\Theta\times\Xi)^{t,J}$ due to (16.8) ;

$P_{l,t,J}$, $P_{u,t,J}$ with respect to $(\Theta\times\Xi)^{t,J}$ according to (16.9);

$P_{l,J}$, $P_{u,J}$ with respect to (16.10)-(16.11) ;

obtaining the extremal distributions for $(\Theta\times\Xi)$ w.r.t. $\mathcal{S}^J$.

#3: (20.1.3)

Solve the lower approximating problem, i.e.

$$\inf_{x\in X}\ f(x) + \int_{\Theta\times\Xi} \rho(\eta,\xi,x)\,dP_{l,J}\ ,$$

and denote a corresponding solution with x^J; if no such x^J exists, then goto #5.

#4: (20.1.4)

Evaluate for $t=1,\dots,t^J$

$$\int_{(\Theta\times\Xi)^{t,J}} \rho(\eta,\xi,x^J)\,dP_{l,J}\ ,\qquad \int_{(\Theta\times\Xi)^{t,J}} \rho(\eta,\xi,x^J)\,dP_{u,J}$$

lower and upper bounds for the expectation of the recourse function with respect to the sub-$\times$-simplices in $\mathcal{S}^J$.

#5: (20.1.5)

If x^J exists, define the set E_ε^J of sub-×-simplices to be split (based on the prescribed tolerance level ε) :

$$E_\varepsilon^J := \{(\Theta\times\Xi)^{t,J} \mid (\Theta\times\Xi)^{t,J}\in \mathcal{S}^J;\ \int_{(\Theta\times\Xi)^{t,J}} \rho(\eta,\xi,x^J)dP_{u,t,J} - \int_{(\Theta\times\Xi)^{t,J}} \rho(\eta,\xi,x^J)dP_{l,t,J} > \varepsilon\} ;$$

else $E_\varepsilon^J := \mathcal{S}^J$

#6: (20.1.6)

If $E_\varepsilon^J = \emptyset$ then STOP ; x^J is sufficiently accurate; i.e. ε-optimal;

else $E_\varepsilon^J \neq \emptyset$:

if x^J exists and the proportion of the longest and the shortest edge does not exceed the tolerance ε^0 :

then split the sub-×-simplices in E_ε^J according to #1' in (19.5) ;

else split the sub-×-simplices in E_ε^J according to #1 in (16.14) .

#7: (20.1.7)

Update $r^{i,J+1}$, $C^{J+1}(\Theta^i)$, $s^{i,J+1}$, $C^{J+1}(\Xi^i)$, $i=1,\dots,I$; t^{J+1} , $\mathcal{S}^{J+1}$

J:=J+1 goto #2 .

Theorem 18.1 ensure that the above procedure ends after a finite number of cycles with an ε-optimal solution, if x^J (J=0,1,2,...) exist and accumulate; otherwise we accept the corresponding approximate values as error bounds.

The success of solving stochastic two-stage problems with the above-outlined algorithm depends strongly on how successful the sequence of deterministic approximation problems in #3 (20.1.3) are solved. These problems are (non)linear mathematical programs for which corresponding methods may be applied. In particular, one has to pay attention to structural properties of these mathematical programs (like (non)convexity, (non)smoothness etc.) for choosing suitable methods. The importance of this feature becomes evident in studying the numerical results achieved within the linear case. Computational experiences and results, including computation times corresponding to above-outlined steps #2-#4 for solving stochastic linear two-stage problems will be reported in chapter VI.

IV An Illustrative Survey of Existing Approaches in Stochastic Two-Stage Programming

As mentioned in chapter I the formulation of stochastic programming problems goes back to the mid 50s (Dantzig, Beale, Tintner). Since then a lot of research activities have resulted in developments of approaches for solving these problems. We intend next to focus on some of existing approaches and outline their main characteristica. The motivation to do this lies in the variety of stochastic two-stage programming problems entailing different and challenging features that have been exploited within the last decades. Some of the approaches discussed below have already been treated in Ermoliev and Wets 1988 [42], and even much more detailed therein than we will do here; however, the achieved progress within stochastic programming in the last five years have raised new ideas.

We adopt the classification introduced in chapter III: Starting with *Approximation Schemes* we report on the works of Kall and Stoyan 1982 [79], Birge and Wets 1986 [14] and Birge and Wets 1989 [16]. Next we pass over to approaches that are assigned to *Stochastic Approximation.* This area has been strongly influenced by the works on stochastic quasigradient (SQG-) methods due to Y. Ermoliev started in the late 60s. Marti 1980 [88] introduced the socalled semi-stochastic approximation aiming to improve the convergence of SQG-methods. Further works that contribute to stochastic approximation are due to Dantzig and Glynn 1990 [25], and Higle and Sen 1991 [56]. Subsequently we discuss cases in which the usage of *Mathematical Programming Techniques* (Nazareth and Wets 1986 [102], 1988 [103]) seems promising. We conclude this section with outlining the concept of scenarios and policy aggregation due to Rockafellar and Wets 1991 [127].

Subsequent approaches have been developed for attacking stochasticity in stochastic programming problems that differ in their applicability with respect to supposed structural properties, as linearity, convexity, separability, sublinearity, absolute continuity. We therefore avoid to outline technical details - as far as possible - and concentrate instead on an illustrative presentation of the main ideas that make these approaches outstanding.

Remark IV.1 (on Notation): Notations referring to the main elements of stochastic optimization problems and their methods coincide with those used in chapters I and III. Nevertheless, for the ease of understanding let us recall the most important: X still stands for a closed, (convex) set of $\mathbf{R}^n$; $\rho(\eta,\xi,x)$ (or $\rho(\xi,x)$, or $\rho(\omega,x)$) represents the integrand (recourse function), where the dimensions of η,ξ,ω are given with K,L,M; P is the underlying probability measure and Θ,Ξ,Ω the support; N stands for a sampling size and J for a certain iteration. In the case of linear recourse problems, W denotes the complete recourse matrix of size (m'×n'); $T(\xi)$ (or $T(\omega)$, or T if deterministic) is the technology matrix of size (m'×n); the socalled induced constraints of the two-stage problem are supposed to be incorporated in the deterministic constraints of the problem, in X respectively. Of course, all integrals are supposed to be well-defined and exist. Additional assumptions on the above functions and sets will be stated explicitly in each section. There is no need to decompose ω into subvectors being stochastically independent; either ω is supposed to be arbitrary, or all components are supposed to be stochastically independent; therefore, using 'I' below has a different meaning than in the previous chapters.

21 Error Bounds for Stochastic Programs with Recourse due to Kall & Stoyan

In considering stochastic programs with recourse Kall and Stoyan 1982 [79] focus on solving

$$\inf_{x\in X} \quad f(x) + \int_{\Omega} \rho(\eta,\xi,x)dP(\eta,\xi) \quad . \tag{21.1}$$

The recourse problem is supposed to be of the following form:

$$\rho(\eta,\xi,x) = \min_y \{\phi(\eta,y) \mid g(\xi,y,x) \leq 0\} \tag{21.2}$$

with $\phi(\eta,y)$ being a continuous, concave-convex saddle function and $(\xi,y)\rightarrow g(\xi,y,x)$ convex for all x in X. The expectation functional reads as :

$$(E\rho)(x) = \int_{\Omega} \rho(\eta,\xi,x)dP(\eta,\xi) \quad . \tag{21.3}$$

The authors approximate the random vector (η,ξ) by (η',ξ') being easier to handle and suppose that the so obtained error $\mid \rho(\eta,\xi,x) - \rho(\eta',\xi',x) \mid$ may be bounded from above according to :

$$\mid \rho(\eta,\xi,x) - \rho(\eta',\xi',x)\mid \leq \|(\eta,\xi) - (\eta',\xi')\| \,.\, \kappa(\eta,\xi,\eta',\xi',x) \quad . \tag{21.4}$$

(η,ξ) and (η',ξ') are treated as random variables that map a common abstract probability space, say $(\mathbf{Z},\mathcal{Z},Z)$ into the (K+L)-dimensional Euclidean space (i.e. (η,ξ): $\zeta\rightarrow(\eta(\zeta),\xi(\zeta))$ and (η',ξ'): $\zeta\rightarrow(\eta'(\zeta),\xi'(\zeta))$) equipped with the Borel algebra and with the induced probability measures P, P' respectively. Assuming square integrability of $\zeta\rightarrow\kappa(\eta(\zeta),\xi(\zeta),\eta'(\zeta),\xi'(\zeta),x)$, $\zeta\rightarrow(\eta(\zeta),\xi(\zeta))$ and of $\zeta\rightarrow(\eta'(\zeta),\xi'(\zeta))$ with respect to Z, we set according to [79] :

$$\|(\eta(\zeta),\xi(\zeta))-(\eta'(\zeta),\xi'(\zeta))\|_2 =$$
$$=[\int_{\mathbf{Z}} <(\eta(\zeta),\xi(\zeta))-(\eta'(\zeta),\xi'(\zeta)),(\eta(\zeta),\xi(\zeta))-(\eta'(\zeta),\xi'(\zeta))>dZ(\zeta)\ ]^{1/2} \tag{21.5.1}$$

$$\kappa^0(x) := [\int_{\mathbf{Z}} [\kappa(\eta(\zeta),\xi(\zeta),\eta'(\zeta),\xi'(\zeta),x)]^2 dZ(\zeta)\ ]^{1/2} \tag{21.5.2}$$

$$(E'\rho)(x) := \int_{\Omega} \rho(\eta',\xi',x)dP'(\eta',\xi') \qquad (21.5.3)$$

With applying Schwarz' inequality to (21.4) Kall and Stoyan obtain an error bound for the expectation functional $(E\rho)(x)$ - (see Theorem 4 in [79]):

$$|(E\rho)(x) - (E'\rho)(x)| \leq \|(\eta(\zeta),\xi(\zeta)) - (\eta'(\zeta),\xi'(\zeta))\|_2 \,.\, \kappa^0(x)\,. \qquad (21.6)$$

Further, supposing that the recourse function $\rho(.,.,x)$ is subdifferentiable and denoting the subgradients at (η_1,ξ_1) with $(\eta^*(\eta_1,\xi_1,x)\ ,\ \xi^*(\eta_1,\xi_1,x))$, the authors apply

$$|\rho(\eta_1,\xi,x) - \rho(\eta_2,\xi,x)| \leq \max_{\eta\in\{\eta_1,\eta_2\}} |\langle(\eta_1 - \eta_2),\ \eta^*(\eta,\xi,x)\rangle| \qquad (21.7.1)$$
$$|\rho(\eta,\xi_1,x) - \rho(\eta,\xi_2,x)| \leq \max_{\xi\in\{\xi_1,\xi_2\}} |\langle(\xi_1 - \xi_2),\ \xi^*(\eta,\xi,x)\rangle| \qquad (21.7.2)$$

obtaining

$$|\rho(\eta_1,\xi_1,x) - \rho(\eta_2,\xi_2,x)| \leq \max_{\eta\in\{\eta_1,\eta_2\}} |\langle(\eta_1 - \eta_2),\eta^*(\eta,\xi_1,x)\rangle| + $$
$$+ \max_{\xi\in\{\xi_1,\xi_2\}} |\langle(\xi_1 - \xi_2),\xi^*(\eta_2,\xi,x)\rangle| \qquad (21.8)$$

and create therefore an expression for

$$\kappa(\eta_1,\xi_1,\eta_2,\xi_2,x) = \max_{\eta\in\{\eta_1,\eta_2\}} |\langle(\eta_1-\eta_2),\eta^*(\eta,\xi_1,x)\rangle| . \|(\eta_1,\xi_1)-(\eta_2,\xi_2)\|^{-1} \qquad (21.9)$$
$$+ \max_{\xi\in\{\xi_1,\xi_2\}} |\langle(\xi_1-\xi_2),\xi^*(\eta_2,\xi,x)\rangle| . \|(\eta_1,\xi_1)-(\eta_2,\xi_2)\|^{-1}$$

for $(\eta_1,\xi_1)\neq(\eta_2,\xi_2)$, and $\kappa(\eta_1,\xi_1,\eta_2,\xi_2,x):=0$ for $(\eta_1,\xi_1)=(\eta_2,\xi_2)$.

As the authors stated, applicability of (21.6) is ensured in stochastic linear programs with complete recourse, where we have square integrability of $\kappa(\eta(\zeta),\xi(\zeta),\eta'(\zeta),\xi'(\zeta),x)$ and boundedness of $\kappa^0(x)$ - due to polyhedrality of the recourse function - yielding

$$\|(\eta(\zeta),\xi(\zeta)) - (\eta'(\zeta),\xi'(\zeta))\|_2 \rightarrow 0 \ \Rightarrow\ (E'\rho)(x) \rightarrow (E\rho)(x) \qquad (21.10)$$

and hence pointwise convergence of the approximate expectation functionals $(E'\rho)(x)$ to $(E\rho)(x)$.

Focusing on $\|(\eta(\zeta),\xi(\zeta))-(\eta'(\zeta),\xi'(\zeta))\|_2$ the authors aim to make the right-hand side in (21.6) as small as possible : denoting a partition of $\mathbf{Z}$ with $\{\ \mathcal{A}^z \mid \mathcal{A}^z \in \mathcal{P}\ ;$

$z=1,...,z^0$ } and its generated σ-algebra with $\mathcal{F}^0$; further, defining the random variable of conditional expectations

$$(\eta^{co}(\zeta),\xi^{co}(\zeta)) = E((\eta(\zeta),\xi(\zeta)) | \mathcal{F}^0) \tag{21.11}$$

Kall and Stoyan 1982 [79] choose

$$(\eta^z,\xi^z) = E\,[(\eta(\zeta),\xi(\zeta)) \mid \zeta \in \mathcal{A}^z] \;\text{ for }\; z=1,...,z^0 \tag{21.12}$$

as $(\eta^{co}(\zeta),\xi^{co}(\zeta))$ minimizes $\|(\eta(\zeta),\xi(\zeta))-(\eta'(\zeta),\xi'(\zeta))\|_2$ subject to all $\mathcal{F}^0$-measurable random variables $(\eta'(\zeta),\xi'(\zeta))$ (see e.g. Bauer 1990 [3]).

In the case of convex recourse functions (then, the random vector reduces to the components of ξ) one gets a lower extremal measure (applying Jensen Inequality with respect to the partition), entailing an approximate problem whose objective minorizes $(E\rho)(x)$ for all x; let x^L denote a solution of this associated lower approximate problem.

For obtaining an upper approximation for (purely convex) stochastic linear programs the authors distinguish 3 cases: i) simple recourse, ii) complete recourse with stochastically independent random components, iii) complete recourse with stochastically dependent random components. Let us shortly review the main ideas:

In the case of *simple recourse* Kall and Stoyan exploit sublinearity and separability of the recourse function and obtain the exact objective value (with respect to x^L), if splitting of the corresponding one-dimensional sublinear recourse function is done at that value $\xi(x^L)$, which splits the function into two linear parts; hence the expectation of $\rho(.,x^L)$ is completely given through the conditional expectations of the 2 new subintervals. The so obtained objective value together with the lower bound according to the above (both referring to x^L), yields an error estimate. In this way, the sequence of refinements is determined through sublinearity of the recourse function at the iterates .

In the case of *complete recourse and stochastically independent components* of ξ within the rectangle Ξ, the authors apply upper extremal measures (induced by Edmundson-Madansky Inequality) with respect to a rectangular refinement of Ξ and solve the associated upper approximate problem obtaining x^U as optimal solutions. Lower and upper approximate problems yield with $(E\rho)(x^L)$ a lower bound and with $(E\rho)(x^U)$ an upper bound for $\inf (E\rho)(x)$. These bounds refer to different first-stage decisions x^L, x^U and hence to different recourse functions $\rho(\eta,\xi,x^L)$, $\rho(\eta,\xi,x^U)$; so, discovering in particular their linear parts of $\rho(\eta,\xi,x^L)$, to ensure efficient partitions, become cumbersome.

Remark 21.1 : This fact had mainly been the motivation for modification outlined in Frauendorfer and Kall 1988 [49]: with adding $x=x^L$ to the upper approximate problem one avoids minimization with respect to x, obtains a less sharp upper bound but with considerably less numerical effort, as the upper approximate problem decomposes into smaller recourse problems with only right-hand side changing. This feature increases the importance of applying duality relations within the refinement procedure. The numerical results - in [49, pp 199] document the experiences with various refinement strategies .

Remark 21.2 : The algorithm outlined in section 20 reduces entirely to that in Frauendorfer and Kall 1988 [49] for the case of stochastic linear programs with complete recourse having stochastically independent random components in the right-hand side and in the technology matrix .

In the case of *complete recourse and stochastic dependent components* Kall and Stoyan replace the upper extremal measure (induced by Edmundson-Madansky Inequality) by that measure which assigns the masses of the rectangles on those vertices subject to which the (convex) recourse function attains its maximum. Therefore contrary to the independent case this discrete distribution is dependent on the first-stage decision and does not exploit linearity. (Computational experiences of that strategy - together with the generalized Edmundson-Madansky Inequality - has been outlined in Frauendorfer 1988 [46].)

22 *Approximation Schemes discussed by Birge and Wets*

Birge and Wets 1986 [14] discuss various approximation schemes that have turned out to be essential for solving stochastic optimization problems; in particular the authors consider

$$\min_{x \in X} \int_{\Xi} \rho(\xi,x)dP(\xi) \tag{22.1}$$

where $\xi \to \rho(\xi,x)$ is assumed to be convex in ξ. We illustrate the main ideas in [14] with the intention to discuss the relations to the results of chapter III. According to [14] we distinguish for our purpose :

i) approximating the recourse function,
ii) discretization of the probability measure throughout conditional expectation,
iii) discrete probability measures with support on extreme points,
iv) extremal probability measures .

i) Approximating the Recourse Function : Focusing on Proposition 3.1 in [14, p.62] we recall that with $\xi \to \rho(\xi,x)$ being convex, subdifferentiable and having in iteration J a finite collection of points $\{\xi^s, s=1,...,s^J\}$ one obtains a lower bound for $(E\rho)(x)$ through

$$(E\rho)(x) \geq \int_{\Xi} \rho^{J,l}(\xi,x)dP(\xi) =: (E\rho^{J,l})(x) \tag{22.2}$$

with

$$\rho^{J,l}(\xi,x) := \max_{s\in\{1,..,s^J\}} [<\xi^{*,s}(x),\xi-\xi^s> + \rho(\xi^s,x)] \tag{22.3}$$

where $\xi^{*,s}(x)$ denotes a subgradient of $\xi \to \rho(\xi,x)$ at ξ^s. The 'max' operator is not necessary for $(E\rho^{J,l})(x)$ to be a lower bound, but to ensure monotonicity of the approximate integrand. As a consequence this implies monotonicity of the associated expectation functional in the case that the collections of support points are successively enlarged. Monotone increasing and pointwise convergent functions entail epi-convergence (see Proposition 4.3), and hence ensure that any accumulation point of approximate solutions solves problem (22.1).

Considering the results in chapter III we may state that in case $\{\xi^s, s=1,\ldots,s^J\}$ represent conditional expectations of subsets (corresponding to refinements of partitions of Ξ) epi-convergence is ensured even without the 'max' operator. This is easily seen in applying Theorem 16.2 to the purely convex case entailing monotonicity of the approximate expectation functionals (but not necessarily of the corresponding approximate integrand).

Remark 22.1: In the above it is supposed that the pointwise convergence is controlled via suitable increasing sequences of points $\{\xi^s, s=1,\ldots,s^J\}$, $J=0,1,\ldots$.

Proposition 3.13 (in [14, p. 65]) provides an upper bound according to

$$(E\rho)(x) \leq \int_{\Xi} \rho^{J,u}(\xi,x)dP(\xi) := (E\rho^{J,u})(x) \tag{22.4}$$

with

$$\rho^{J,u}(\xi,x) = \inf \{ \Sigma \lambda^s.\rho(\xi^s,x) \mid \xi = \Sigma \lambda^s.\xi^s, \Sigma \lambda^s = 1, \lambda^s \geq 0, s=1,\ldots,s^J \} \tag{22.5}$$

where again $\{\xi^s, s=1,\ldots,s^J\}$ is an arbitrary collection of points in Ξ and the 'inf' operator is needed for enforcing monotonicity of the upper approximate in the case that the collection of points is enlarged.

Considering the constructions of upper approximates in chapter III and the upper bound in (22.4) we conclude: within our special case we work with a ×-simplex Ξ with $\{\xi^s, s=1,\ldots,s^J\}$ representing vertices of Ξ; the upper bounds are determined through (generalized) barycentric weights with respect to Ξ, avoids minimization with respect to all possible convex combinations (that would yield a sharper upper bound computationally more expensive).

ii) Discretization of Probability Measures throughout Conditional Expectations : Birge and Wets 1986 [14] focus on Jensen Inequality with respect to partitioning the rectangle Ξ. As refinement strategy it is suggested to exploit linearity in considering inhowfar the dual variables of the recourse problems (associated with the vertices of subrectangles) differ most with respect to a predefined norm. This strategy has been adopted in Frauendorfer and Kall 1988 [49] and therein compared to the purely convex counterpart of (19.5).

iii) Discrete Probability Measures with Support on Extreme Points: In that part Birge and Wets use a very general approach for majorizing expectation functionals with respect to any compact, convex support: In denoting the set of extreme points by ext-Ξ and $\mathcal{E}$ as the corresponding Borel field, the authors get by convexity of $\xi \to \rho(\xi,x)$

$$(E\rho)(x) \leq \int_{\text{ext-}\Xi} \rho(\xi,x)d\lambda(\xi) \tag{22.6}$$

where λ is supposed to be a probability measure on (ext-Ξ, $\mathcal{E}$).

Considering the results of section 14 we conclude that extending (22.6) in this generality for upper approximations of saddle functions necessitates to consider the entire boundary of Ξ (instead of ext-Ξ) and its Borel field.

iv) Extremal Probability Measures: Birge and Wets 1986 [14, pp 81-90] consider a set of probability measures Δ - supposed to contain P - being given through a system of equalities and inequalities

$$\Delta := \{ Q \mid \int_{\Xi} v_i(\xi)dQ \leq \alpha_i \ (i=1,\ldots,I') \wedge \int_{\Xi} v_i(\xi)dQ = \alpha_i \ (i=I'+1,\ldots,I'') \} \tag{22.7}$$

where $v_i(\xi)$ are supposed to be continuous functions on Ξ for i=1,...,I''; the set in (22.7) induces the following pair of generalized moment problems :

$$\inf_{Q\in\Delta} \int_{\Xi} \rho(\xi,x)dQ(\xi) \quad , \quad \sup_{Q\in\Delta} \int_{\Xi} \rho(\xi,x)dQ(\xi) \quad . \tag{22.8}$$

A characterization of the solutions in (22.8) is given with Theorem 6.9 in [14, p.82] being recalled within our context as

Proposition 22.1 : Suppose Ξ is compact. Then the set Δ is convex and compact (with respect to the weak topology), and Δ=cl(co(ext-Δ)). Assuming that for any x in X, $\xi \to \rho(\xi,x)$ is continuous relative to Ξ, then (22.8) are solved at an extreme point (i.e. extremal measure) of Δ. Moreover, the extremal measures of Δ are precisely those having finite support $\{\xi^1,\ldots,\xi^s\}$ with $s \leq I''+1$ such that

$$\begin{pmatrix} v_1(\xi^1) \\ v_2(\xi^1) \\ \cdot \\ v_{I''}(\xi^1) \\ 1 \end{pmatrix}, \begin{pmatrix} v_1(\xi^2) \\ v_2(\xi^2) \\ \cdot \\ v_{I''}(\xi^2) \\ 1 \end{pmatrix}, \ldots, \begin{pmatrix} v_1(\xi^s) \\ v_2(\xi^s) \\ \cdot \\ v_{I''}(\xi^s) \\ 1 \end{pmatrix} \tag{22.9}$$

are linearly independent.

In section 14, the moment problems (22.8) have been solved for the case $\xi \to \rho(\xi,x)$ being a continuous saddle function (relative to a $\times$-simplex Ξ) and $v_i(\xi)$ (i=1,...,I'') being multilinear, bilinear respectively.

23 Sublinear Bounding Technique due to Birge and Wets

Birge and Wets 1989 [16] aim to solve stochastic linear programs with complete recourse of the following form :

$$\min \langle c,x\rangle + \int_{\Xi} \rho(\xi,\chi)dP(\xi) \tag{23.1.1}$$

$$\text{s.t.} \quad Ax = a \tag{23.1.2}$$

$$Tx - \chi = 0 \tag{23.1.3}$$

$$x \geq 0$$

with

$$\rho(\xi,\chi) = \min \{\langle q,y\rangle \mid Wy = \xi - \chi \ , \ y \geq 0\} \ ; \tag{23.2}$$

herewith, technology matrix and objective coefficients are supposed to be deterministic; $\xi,\chi \in \mathbf{R}^{m'}$ and $\rho(\xi,\chi):\mathbf{R}^{m'} \times \mathbf{R}^{m'} \to \mathbf{R}$ throughout this section. Exploiting sublinearity of the recourse function with respect to $(\xi - \chi)$ the authors concentrate on solving 2m' linear recourse problems

$$\min \langle q,y\rangle \qquad\qquad \min \langle q,y\rangle \tag{23.3.1}$$

$$\text{s.t.} \ Wy = +d_i \qquad\qquad \text{s.t.} \ Wy = -d_i \tag{23.3.2}$$

$$y \geq 0 \qquad\qquad y \geq 0$$

where $i=1,...,m'$ and $\{d_1,...,d_{m'}\}$ is a basis for $\mathbf{R}^{m'}$. Thus, the collection $D=\{d_1,...,d_{m'}, -d_1, ...,-d_{m'}\}$ allows to represent any element of $\mathbf{R}^{m'}$ through a positive linearcombination. For the ease of exposition let $i=1,...,m',m'+1,...,2m'$ and define $d_{m'+1}=-d_1, ..., d_{m'+m'}=-d_{m'}$; further we denote with $I(\xi-\chi)$ those indices of linearly independent columns of D subject to which $\xi-\chi$ may be written as a positive linearcombination; let $D(\xi-\chi)$ denote the regular matrix associated with $I(\xi-\chi)$. Having $\xi-\chi$ as right-hand side and denoting with $[D(\xi-\chi)^{-1}]_{i,.}$ that row of $D(\xi-\chi)^{-1}$ that corresponds to d_i, we obtain

$$(\xi-\chi) = \Sigma_{i\in I(\xi-\chi)} <[D(\xi-\chi)^{-1}]_{i,.}, \xi-\chi > . d_i \tag{23.4}$$

and as a consequence an upper approximation for $\rho(\xi,\chi)$ through :

$$\rho(\xi,\chi) \leq \Sigma_{i\in I(\xi-\chi)} <[D(\xi-\chi)^{-1}]_{i,.}, \xi-\chi > . <q,y^{opt}(d_i)> := \rho_D(\xi,\chi) . \tag{23.5}$$

$y^{opt}(d_i)$ represents the optimal solution of the recourse problem having d_i as right-hand side in (23.3.2). As collection D has been chosen arbitrarily one may conclude that a finite set $\mathcal{D}$ of such collections improves the upper approximation through

$$\rho(\xi,\chi) \leq \inf_{D\in\mathcal{D}} \rho_D(\xi,\chi) = \mathrm{co}_{D\in\mathcal{D}} [\rho_D(.,\chi)](\xi) \tag{23.6}$$

getting after integration

$$\int_\Xi \rho(\xi,\chi)dP(\xi) \leq \int_\Xi [\inf_{D\in\mathcal{D}} \rho_D(\xi,\chi)] \, dP(\xi) =$$

$$= \int_\Xi \mathrm{co}_{D\in\mathcal{D}} [\rho_D(.,\chi)](\xi) \, dP(\xi) . \tag{23.7}$$

$\mathrm{co}_{D\in\mathcal{D}} [\rho_D(.,\chi)](\xi)$ is obtained by taking the convex hull of $\{\mathrm{epi}\ \rho_D(.,\chi) \mid D\in\mathcal{D}\}$.

If $\mathcal{D}$ includes all linear bases of W then (23.6) and (23.7) holds as equality. Due to the fact that (23.7) becomes computationally cumbersome - minimization is required inside the integral - the authors suggest to reverse the infimum and integration, obtaining (due to Proposition 3.1 in [16]):

$$\int_\Xi \rho(\xi,\chi)dP(\xi) \leq \int_\Xi [\inf_{D\in\mathcal{D}} \rho_D(\xi,\chi)] \, dP(\xi) = \tag{23.8.1}$$

$$= \int_{\Xi} co_{D\in\mathcal{B}} [\rho_D(.,\chi)](\xi)\, dP(\xi) \le co_{D\in\mathcal{B}} [\int_{\Xi} \rho_D(\xi,\chi) dP(\xi)] =: co_{D\in\mathcal{B}} (E\rho_D)(\chi). \quad (23.8.2)$$

In (23.8.2) equality holds only in very special cases (even if $\mathcal{B}$ includes all bases of W, '=' cannot be guaranteed).

For evaluating an approximate solution and an upper bound for the original problem (23.1) Birge and Wets state as approximate problem:

$$\min \; \langle c,x\rangle + co_{D\in\mathcal{B}} (E\rho_D)(\chi) \quad (23.9.1)$$

$$\text{s.t.} \quad Ax = a \quad (23.9.2)$$

$$Tx-\chi = 0 \quad (23.9.3)$$

$$x\ge 0 \; .$$

Avoiding to take the convex hull of a set of functions the authors apply Theorem 16.5 in Rockafellar 1970 [118] and pass over to the dual program of (23.9). Denoting with ρ_D* the conjugate of ρ_D yields

$$[co_{D\in\mathcal{B}} (E\rho_D)]^* = sup_{D\in\mathcal{B}} (E\rho_D)^* \; . \quad (23.10)$$

The dual of (23.9) may be written according to Proposition 4.1. in Birge and Wets 1989 [16] as

$$\max \; \langle v,a\rangle - sup_{D\in\mathcal{B}} (E\rho_D)^*(-w) \quad (23.11.1)$$

$$\text{s.t.} \quad vA + wT \le c \quad (23.11.2)$$

where v,w denote the dual variables of (23.9).

For getting expressions for ρ_D* one requires the marginal distribution function of the transformed random variables $\langle [D(\xi-\chi)^{-1}]_{i,.},\xi-\chi\rangle$, being available e.g. for normally and independently distributed random variables ξ [16, pp 139].

Several generation techniques for $\mathcal{B}$ as well as an algorithmic description having been applied for test problems are given in Birge and Wets 1989 [16]. Due to construction the outlined algorithm amounts to solve O(m') linear programs per iteration. Therefore, the sublinear bounding technique depend linearly on the dimension of the underlying probability space.

Considering the stochastic convex optimization problem

$$\inf_{x\in X} \ (E\rho)(x) = \inf_{x\in X} \int_{\Xi} \rho(\xi,x)\, dP(\xi) \qquad (24.1)$$

we get due to convexity of $(E\rho)(x)$

$$(E\rho)(x) - (E\rho)(x') \geq \langle (E\rho)_x(x), x-x' \rangle \qquad (24.2)$$

where $(E\rho)_x(x)$ denotes a subgradient of $(E\rho)$ at x. *Stochastic quasigradient (SQG) techniques* have been initiated by Y. Ermoliev in the mid 60s; for the history of these methods see e.g. Wets 1983 [142]. In this approach the sequence of approximate solutions $\{x^J;\ J=0,1,....\}$ for (24.1) are constructed by using statistic estimates, based on the iterates $\{x^0,...,x^{J-1}\}$ obtained within the optimization process. These statistic estimates represent (essentially) subgradients, being random variables denoted x^* and close to $(E\rho)_x$ 'in average'; in particular for iterate x^J one may write

$$E(x^{*,J} \mid x^0,...,x^J) = (E\rho)_x (x^J) + \alpha^J \ . \qquad (24.3)$$

Convergence to an optimal solution necessitates $\|\alpha^J\| \rightarrow 0$; (if $\alpha^J \neq 0$ then $x^{*,J}$ is called a *stochastic quasigradient*; in the case $\alpha^J=0$ then $x^{*,J}$ is a *stochastic subgradient.*) In literature different special rules for computing $x^{*,J}$ have been developed; herein, we restrict ourselves to the common one

$$x^{*,J} = (E\rho)_x (\xi^J, x^J) \qquad (24.4)$$

or more generally

$$x^{*,J} = (1/N^J).\ \Sigma_s\, (E\rho)_x (\xi^{J,s}, x^J) \qquad (24.5)$$

where ξ^J, $\xi^{J,s}$ $(s=1,...,N^J)$ represent realizations in iteration J yielding subgradients $\rho_x(\xi^J,x)$, $\rho_x(\xi^{J,s},x)$ of $x \rightarrow \rho(\xi^J,x)$, $x \rightarrow \rho(\xi^{J,s},x)$ $(s=1,...,N^J)$. Having an estimate $x^{*,J}$ for the subgradient of $(E\rho)$ at x^J the next step is to apply the procedure that yields x^{J+1}; applying, for example, the projection method one gets

$$x^{J+1} = \Pi_X \; [x^J - \gamma^J.x^{*,J}] \tag{24.6}$$

with Π_X as projection of $(x^J-\gamma^J.x^{*,J})$ onto X. Under appropriate assumptions like, compactness of the solution set and uniform integrability of the stochastic quasigradients (in second order), then it has been proven that in particular the projection method sketched out above converges to an optimal solution (with probability one) for positive step-sizes $\{\gamma^J, J=1,2,...\}$ which satisfy the following type of condition (with probability one) :

$$\Sigma_J \, (\gamma^J)^2 < \infty \, , \;\; \Sigma_J \, \gamma^J = \infty \; . \tag{24.7}$$

Condition (24.7) enforces that the chosen sequence of step sizes converge neither too fast nor too slow to 0. (For the details and for generalized versions of the above technique we refer to Ermoliev 1983 [40], 1988 [41].)

SQG methods mainly concentrate on approximating x^{opt}. P as well as the objective function ψ are approximated rather roughly (in an individual iteration) and nonmonotonously. This is due to the fact that SQG methods have been designed mainly to solve optimization problems of rather complex type. The convergence rate of these methods is rather low due to its probabilistic nature and given with

$$E\,(\|x^J-x^{opt}\|^2) = O(1/J^{\beta}) \qquad (0<\beta\leq 1)\; . \tag{24.8}$$

This increases the importance of stopping criterions subject to which sufficiently accurate approximates are realized. Stopping criterions rely mainly on statistics of objective values at the iterates, on statistics of step sizes or on statistics of subgradients indicating 'how likely' optimality conditions are satisfied. Therefore, the entire computational effort and the success of the application depends strongly on how efficient objective values, subgradients, step sizes are evaluated and stopping criteria are specified. There exist several works on attacking these criterions; to mention only some: Ruszczyński and Syski 1986 [134], Ruszczyński 1987 [133], Pflug 1988 [104], Gaivoronski 1988 [50], Higle and Sen 1991 [57].

A. Gaivoronski designed an impressive software package - named SQG-PC - that solves problems of the form (24.1) with employing different SQG techniques and providing user-friendly utilities for controlling the above stated criterions. For details about this package consult its User's Guide (Gaivoronski 1988 [51]).

Marti 1980 [88] attacks nonmonotonicity and improves the convergence behavior of stochastic quasigradient techniques with introducing the so-called semi-stochastic approximation. The essential idea is to replace stochastic quasi-gradients by deterministic descent directions in some (infinitely many) iterations. Valuing the difference $[T(\xi).x-h(\xi)]$ by an utility function Marti considers stochastic optimization problems of the form

$$\inf_{x\in X}\ (Eu)(x) := \int_{\Xi} u(T(\xi).x - h(\xi))\, dP(\xi) \tag{25.1}$$

where $u:\mathbf{R}^{m'}\to\mathbf{R}$ is supposed to be a convex, (sub)differentiable loss function. Having a certain iterate, say $x'\in X$, the problem is to compute a suitable descent direction. For illustrative purposes think of P being finite and discrete, in particular having $\{\xi^1,\dots,\xi^N\}$ as support, and suppose that we have two points, say x' and x" both in X; then, two collections of points are given according to:

$$\{T(\xi^1).x' - h(\xi^1), \dots , T(\xi^N).x' - h(\xi^N)\} \tag{25.2.1}$$
$$\{T(\xi^1).x'' - h(\xi^1), \dots ,T(\xi^N).x'' - h(\xi^N)\}\ . \tag{25.2.2}$$

If we further have

$$\begin{aligned} &\mathrm{co}\{T(\xi^1).x' - h(\xi^1), \dots , T(\xi^N).x' - h(\xi^N)\} \subset \\ &\quad\subset \mathrm{co}\{T(\xi^1).x'' - h(\xi^1), \dots ,T(\xi^N).x'' - h(\xi^N)\} \end{aligned} \tag{25.2.3}$$

and coincidence in expectation

$$\int_{\Xi} (T(\xi).x' - h(\xi))\, dP(\xi) = \int_{\Xi} (T(\xi).x'' - h(\xi))\, dP(\xi) \tag{25.2.4}$$

then one may conclude that d=x"-x' is a feasible deterministic descent direction for any strictly convex function u(.). Exploiting and distinguishing structural properties of the loss function and of the distribution function Marti has introduced more general counterparts to properties (25.2.1)-(25.2.4) called *stochastic dominance conditions* in Marti 1988 [90] (and references therein). Sophisticated conditions for verifying these properties as well as tools that allow to determine an appropriate x' via solving systems of linear equations or specially structured optimization problems have been worked out in Marti 1985 [89], Marti and Fuchs 1986 [92]. Based on these features the deterministic descent direction (in the case of existence) is obtained without any information of derivatives of (Eu)(x) and is stable with respect to a certain class of loss functions. In the case that no deterministic descent direction exists then x' is denoted *efficient* solution with respect to X and with respect to the above mentioned class of utility function. If x' is accepted as sufficient accurate solution for problem (25.1) or as a solution that hedges against perturbations in the loss function then we are done. To the contrary, if we focus on verifying optimality of x' (at least in a statistical sense) or on improving the solution then one has to pass over to e.g. stochastic quasigradient techniques. Due to the above features, it is close to rely on feasible deterministic descent directions whenever possible. This yields the socalled *semi-stochastic approximation* procedure being stated in accordance with Marti and Fuchs 1986 [91] :

$$x^{J+1} = \begin{cases} \Pi_X \, [x^J + \gamma^J . d^J] & \text{if } J \in \mathbf{N}^1 \\ \Pi_X \, [x^J - \gamma^J . x^{*,J}] & \text{if } J \in \mathbf{N}^2 \end{cases} \tag{25.3}$$

where $\mathbf{N}^1$, $\mathbf{N}^2$ denote subsets of all integers $\mathbf{N}$ with $\mathbf{N}=\mathbf{N}^1 \cup \mathbf{N}^2$; the sequence $\mathbf{N}^1$ represent all those iterations for which a deterministic descent direction is computable; hence $\mathbf{N}^2$ covers all other iterations performing a stochastic step according to (24.6). In Marti 1980 [88] convergence (with probability one) of the procedure in (25.3) has been proven under appropriate assumptions for positive step sizes γ^J satisfying

$$\Sigma_{J \in \mathbf{N}} (\gamma^J)^2 < \infty , \; \Sigma_{J \in \mathbf{N}^2} \gamma^J = \infty . \tag{25.4}$$

Concerning the speed of convergence Marti and Fuchs 1986 [91, Theorem 5.1] derived that

$$E \, (\|x^J - x^{opt}\|^2) = O(1/J^\beta) \quad (0<\beta\leq 2-1/r\,) \tag{25.5}$$

supposing (among other assumptions) to take stochastic steps at a decreasing rate; more precisely $\mathbf{N}^2 = \{j^r \mid j=1,2,...\}$ for some integer r. Observe that for r=1 (25.5) covers (24.8). Step size rules for (25.3) are thoroughly outlined in Marti and Plöchinger 1990 [93,94]: Optimizing the sequence of step sizes with respect to error bounds - being given through recurrence relations - the authors succeeded in deriving recursions for optimal step sizes. Under certain circumstances these step sizes entail a linear convergence rate of the corresponding semi-stochastic procedure.

26 Benders' Decomposition with Importance Sampling due to Dantzig & Glynn

Let us think of solving approximately a given stochastic programming problem by means of sampling the distribution function P with a sampling size N^{MC} (MC stands for crude Monte Carlo sampling). It is well known from the central limit theorem that the error of the obtained estimate depends on the variance and that the corresponding convergence rate being rather slow of order $O([N^{MC}]^{-1/2})$. The technique of *importance sampling* is a possibility to strongly reduce the variance of an estimate (and so the error). The essential idea in importance sampling lies in focusing on those parts of the support that matter more. This necessitates to exploit available information of integrands. Things become a bit harder within stochastic programming as the integrand usually is not known explicitly and additionally equipped with the socalled first-stage solution x as parameter. Dantzig and Glynn 1990 [25] proposed a way to overcome this difficulty and embed importance sampling within Benders' decomposition for exploiting structural properties of the problem at hand. In accordance with [25] we consider stochastic linear programs with recourse of the form

$$\min \langle c,x\rangle + \int_{\Omega} \rho(\omega,x)dP(\omega) \tag{26.1.1}$$

$$\text{s.t.} \quad Ax = a \tag{26.1.2}$$

$$x \geq 0$$

with

$$\rho(\omega,x) = \min \{\langle q(\omega),y\rangle \mid Wy= h(\omega)-T(\omega)x ,y\geq 0\} \tag{26.1.3}$$

and $\omega \in \mathbf{R}^M$ with components supposed to be stochastically independent.

Remark 26.1: As in the following approach there is no necessity to distinguish between the concave and convex part of $\rho(.,x)$ we use ω instead of (η,ξ). Further, in [25] the recourse matrix W is allowed to be random, too. In this part we only relax completeness of W and keep W deterministic.

We recall that for finite support Ω, (26.1) becomes a LP

$$\min \ <c,x> + \Sigma_{\omega\in\Omega} <q(\omega),y(\omega)>.p(\omega) \tag{26.2.1}$$

$$\begin{aligned} \text{s.t.} \quad & Ax = a && (26.2.2)\\ & T(\omega)x + Wy(\omega) = h(\omega) \quad \forall\ \omega\in\Omega && (26.2.3)\\ & x,\ y(\omega) \geq 0 \ . \end{aligned}$$

Applying Benders' decomposition (see Benders 1962 [6]) we get as master problem

MASTER

$$\min \ <c,x> \ + \ \theta \tag{26.3.1}$$

$$\begin{aligned} \text{s.t.} \quad & Ax = a && (26.3.2)\\ & <g_j,x> + \delta_j.\theta \geq \gamma_j \quad (j=1,\dots,J) && (26.3.3)\\ & x \geq 0 && (26.3.4) \end{aligned}$$

and (x^J, θ^J) as optimal solution. The subproblems read as follows $\forall\ \omega\in\Omega$:

SUBPROBLEM

$$\begin{aligned} & \min \ <q(\omega),y> && (26.4.1)\\ & \text{s.t.} \quad Wy = h(\omega) - T(\omega)x^J && (26.4.2)\\ & \qquad y \geq 0 && (26.4.3) \end{aligned}$$

with $y(\omega,x^J)$, $\pi(\omega,x^J)$ denoting an optimal primal, dual solution.

Constraints (26.3.3) represent the j-th cut obtained in iteration j from the subproblems through

$$\begin{aligned} g_j &:= \Sigma_{\omega\in\Omega} [\pi(\omega,x^j).T(\omega)].p(\omega) && (26.5.1)\\ \gamma_j &:= \Sigma_{\omega\in\Omega} <\pi(\omega,x^j), h(\omega)>.p(\omega) && (26.5.2) \end{aligned}$$

in the case that all subproblems (26.4) are feasible with respect to x^j; this cut is denoted optimality cut with $\delta_j:=1$. In the case that in iteration j some subproblem ω is infeasible with respect to x^j then $\delta_j:=0$; with $\pi(\omega,x^j)$ denoting the optimal dual solution of the infeasible subproblem (26.4) extended with artificial variables (infeasibility phase), the socalled infeasibility cut is given with :

$$\begin{aligned} g_j &:= [\pi(\omega,x^j).T(\omega)] && (26.6.1)\\ \gamma_j &:= <\pi(\omega,x^j), h(\omega)> \ . && (26.6.2) \end{aligned}$$

The method terminates if all subproblems are feasible and $\theta^J \geq \gamma_J - < g_J, x^J >$.

Obviously

$$(E\rho)(x^J) = \Sigma_{\omega \in \Omega} <q(\omega),y(\omega,x^J)>.p(\omega) \qquad (26.7)$$

or more generally

$$(E\rho)(x^J) = \int_{\Omega} <q(\omega),y(\omega,x^J)>dP(\omega) \ . \qquad (26.8)$$

Above integrals are computable if the support of P is not too large. In the case that Ω covers, say, billion of data, one has to sample Ω obtaining Ω^{MC} as support and P^{MC} as probability distribution. This yields an estimate $(E^{MC}\rho)(x^J)$ for $(E\rho)(x^J)$ where

$$(E^{MC}\rho)(x^J) := \int_{\Omega^{MC}} <q(\omega),y(\omega,x^J)>dP^{MC}(\omega) =$$
$$= \Sigma_{\omega \in \Omega^{MC}} <q(\omega),y(\omega,x^J)>.p^{MC}(\omega) \ . \qquad (26.9)$$

The smaller the variance of $(E^{MC}\rho)(x^J)$ the smaller the error and the better the solution x^{opt} of the corresponding deterministic equivalent problem (26.2) being solved with Benders' decomposition. To reduce the variance of the estimate - in each iteration J - one may either enlarge the sample size or apply the importance sampling technique. As increasing the sample size N^{MC} is certainly limited by the dimensions of the problem (26.1) and entails only modest improvements due to the convergence rate $O([N^{MC}]^{-1/2})$, Dantzig and Glynn 1990 [25] recommend importance sampling: below, we illustrate the way, the authors derive an importance sampling distribution $P^{IS,J}$ with corresponding support $\Omega^{IS,J}$, and sampling size $N^{IS,J}$ (J as superscript emphasizes the dependence on x^J and hence on iteration J).

It is well known that to what extent the variance can be reduced depends mainly on inhowfar the probability mass function $p^{IS,J}(.)$ satisfies

$$p^{IS,J}(\omega) \sim <q(\omega),y(\omega,x^J)>.p(\omega) = q^J(\omega).p(\omega) \qquad (26.10)$$

with $q^J(\omega) := <q(\omega),y(\omega,x^J)>$ (see e.g. Davis and Rabinowitz 1975 [27]). Due to the complexity of the problem it is not affordable to concentrate on the information on the right-hand side of (26.10). Instead the authors intend to achieve for the importance sampling distribution

$$p^{IS,J}(\omega) \sim [\sum_{i=1}^{M} q_i^J(\omega_i)].p(\omega) \tag{26.11}$$

where $\Sigma_i\ q_i^J(\omega_i)$ is seen as approximation for $q^J(\omega)$. The costs $q_i^J(\omega_i)$ are supposed to be given through the slopes of the recourse function between lattice points. More precisely, $q_i^J(\omega_i)$ are determined through the costs $q^J(\tau)$ of a socalled base case τ, for i=1,...,M, via :

$$[q^J(\tau_1,...,\tau_{i-1}, \omega_i, \tau_{i+1},...,\tau_M) - q^J(\tau_1,...,\tau_{i-1}, \tau_i, \tau_{i+1},...,\tau_M)]\ . \tag{26.12}$$

The differences in (26.12) are obtained through evaluating the recourse function on a relatively small set of lattice points along M coordinate directions. In order that $p^{IS,J}(\omega)$ gets a probability mass function and (26.11) is satisfied, Dantzig and Glynn define for ω in Ω (supposed to be finite) :

$$p^{IS,J}(\omega) := [\Sigma_\omega [\Sigma_i q_i^J(\omega_i)].p(\omega)]^{-1} . [\Sigma_i q_i^J(\omega_i)] . p(\omega) = \tag{26.13}$$
$$[\Sigma_i Eq_i^J]^{-1} . \Sigma_i \{[q_i^J(\omega_i).p_i(\omega_i)].\Pi_{i' \neq i}\, p_{i'}(\omega_{i'})\},$$

where Eq_i^J denote the expected costs with respect to the i-th component of ω and with respect to the additive approximation in (26.11). These M expectations are supposed to be obtained through sampling the marginal distributions corresponding to ω_i. (The additive approximates for the costs are already available through the chosen lattice points.)

For determining an importance sampling of size $N^{IS,J}$, the terms (i=1,...,M)

$$\{[q_i^J(\omega_i.p_i(\omega_i)]\ .\Pi_{i' \neq i}\, p_{i'}(\omega_{i'})\} \tag{26.14}$$

in (26.13) motivate to sample M times with sampling sizes $N_i^{IS,J}$ (i=1,...,M) for which $\Sigma_i\ N_i^{IS,J}=N^{IS,J}$: the i-th group of realizations arises through sampling $\omega_{i'}$ for $i' \neq i$ according to the corresponding marginal distribution of P and through sampling ω_i according to

$$p_i^{IS,J}(\omega_i) = [q_i^J(\omega_i).p_i(\omega_i)] . [Eq_i^J]^{-1}\ ; \tag{26.15}$$

note that the expression in (26.14) has been necessarily standardized. Therefore $p_i^{IS,J}(\omega_i)$ may be seen as probability mass function of the importance distribution $P_i^{IS,J}$ (for the i-th component of ω). Due to the standardization in (26.14) the desired importance distribution is then given with the probability mass function

$$p^{IS,J}(\omega) := [\Sigma_i Eq_i^J]^{-1}. \Sigma_i [Eq_i^J].\{ [p_i^{IS,J}(\omega_i)].\Pi_{i' \neq i} p_{i'}(\omega_{i'}) \} \qquad (26.16)$$

being proportional to $[q_i^J(\omega_i).p_i(\omega_i)]$. This yields the importance distribution $P^{IS,J}$ and an associated importance sampling with support denoted $\Omega^{IS,J}$.

Remark 26.2: Obviously one has to ensure $[\Sigma_\omega q_i^J(\omega_i).p(\omega)]^{-1} \neq 0$; this is suggested to be achieved with a suitable choice of the base case; e.g. with choosing τ to be that realization with the smallest (unique) second stage costs.

Having solved the subproblems (26.4) given through the importance sample one may expect a considerable reduction in the variance of the estimate

$$(E^{IS,J}\rho)(x^J) := \int_{\Omega^{IS,J}} \langle q(\omega), y(\omega, x^J)\rangle dP^{IS,J}(\omega) . \qquad (26.17)$$

If $\theta^J \geq \gamma_J - \langle g_J, x^J \rangle$ then x^J is accepted as approximate optimal solution; otherwise the cut given through the subproblems associated with the importance sample is transmitted to the master problem (26.3) being solved next, yielding a new iterate x^{J+1}.

Due to construction Dantzig and Glynn call the above-outlined method an *additive importance sampling method.* The achieved variance reduction of (26.17) depends on the extent of which (26.10) is satisfied. Applying this method to an application problem within electric power systems Dantzig et al. 1989 [26] claim a surprising variance reduction of 1:20000 in comparing *crude* versus *importance* sampling. In a next step Infanger 1991 [63] has extended the above outlined technique: with introducing the socalled probabilistic cuts Infanger determines estimates not only for the expected costs of the second stage but also for the cuts and their right-hand sides with respect to the same importance sampling. Considering test problems that allow to be solved as deterministic equivalent problems Infanger 1991 [63] reports on promising first results based on comparing the exact solutions with the obtained approximates.

Higle and Sen 1991 [56] present the socalled *stochastic decomposition* with the intention to combine elements of Benders' decomposition technique (of decomposition methods, in general) and elements of stochastic approximation. The authors consider stochastic linear programs with complete recourse

$$\min_{x\in X} \; \langle c,x\rangle + \int_{\Xi} \rho(\xi,x)dP(\xi) \tag{27.1.1}$$

with

$$\rho(\xi,x) = \min \{\langle q,y\rangle \mid Wy = \xi - T.x,\; y\geq 0\} . \tag{27.1.2}$$

X and Ξ are supposed to be compact; the dual feasible region associated with the recourse problem is supposed to be a compact convex polyhedral set, for all $x\in X$; apparently, cost coefficients and technology matrix are deterministic. We have to stress that deterministic cost coefficients imply a deterministic (nonempty) dual feasible recourse region, supposed to be characterizable by a set V, representing finitely many extreme points.

Thinking in the terminology of Benders' decomposition (due to the preceding section) the authors suggest to solve only one subproblem per iteration J, and J-1 strongly relaxed recourse problems determined through an iterate x^J and a randomly generated sample ξ^J (w.r.t. P). Due to the fact that the subproblems differ from iteration to iteration with respect to x <u>and</u> ξ, this requires to update the cuts (obtained in iteration 1,...,J-1) in the master problem. For the ease of exposition we start with outlining the algorithmic description of the socalled *basic method* according to Higle and Sen 1991 [56]:

#0 : $J:=0$, $V^0=V^J:=\emptyset$, $\xi^0=\xi^J:=E(\xi)$ and choose x^1 as solution of the initial master: $\min\{\langle c,x\rangle+\rho(\xi^J,x) \mid x\in X\}$;

#1 : $J:=J+1$; randomly generate ξ^J according to P;

#2 : solve the subproblem : $\min \{\langle q,y\rangle \mid Wy=\xi^J-T.x^J,\; y\geq 0\}$ to obtain an optimal dual solution $\pi(\xi^J,x^J)$ and create $V^J:= V^{J-1}\cup\{\pi(\xi^J,x^J)\}$;

#3 : generate the J-th cut $\langle g_{J,J},x\rangle+\theta \geq \gamma_{J,J}$ to be added to the master problem with: $g_{J,J} :=(1/J).\Sigma_j\, [\pi(\xi^j,x^J).T]$ and $\gamma_{J,J} := (1/J)\, .\Sigma_j\, [\langle\pi(\xi^j,x^J),\xi^j\rangle$ whereby $\pi(\xi^j,x^J)$ represents a solution of $\max\{\langle\pi,\, \xi^j-T.x^J\rangle \mid \pi\in V^J\}$ for $j=1,...,J-1$;

#4 : update previously generated cuts through
$g_{j,J} := g_{j,J-1}.(J-1)/J;\ \gamma_{J,J} := \gamma_{j,J-1}.(J-1)/J$ for j=1,...,J-1 ;

#5 : solve the J-th master : $\min \{<c,x>+\theta \mid <g_{j,J},x>+\theta \geq \gamma_{j,J} ;\ j=1,\ldots,J,\ x\in X\}$ to obtain x^{J+1} as solution. Repeat from #1 .

Let us now roughly discuss the above procedure regarding its differences to Benders' decomposition technique on the basis of J realizations up to iteration J:
In iteration J of Benders' decomposition technique we have that the J-th cut is determined through taking the expectation of $[\pi(\xi,x^J).T]$ and $<\pi(\xi,x^J),\xi>$ with respect to all J realizations and solving all J associated subproblems. In step #2 of the basic method Higle and Sen solve only the subproblem that corresponds to the last randomly generated realization ξ^J; for the other J-1 realizations the authors solve J-1 strongly relaxed subproblems, obtained through substituting V^J for V.
Prior to iteration J of Benders' decomposition, in particular the j'-cut refers also to J realizations and the associated subproblems and to the iterate $x^{j'}$. In #4 Higle and Sen obtain the following formulae for the j'-th cut in iteration J - according to the proposed update:

$$g_{j',J} := (1/J) \sum_{j=1}^{j'} [\pi(\xi^j,x^{j'}).T] \tag{27.2.1}$$

$$\gamma_{j',J} := (1/J) \sum_{i=1}^{j'} <\pi(\xi^j,x^{j'}),\xi^j> ; \tag{27.2.2}$$

hence the derived cuts serve as statistical estimates of supports for $(E\rho)(x)$. (27.2) implies that the slopes of the cuts tend to 0. Thus, if the sequence of iterates have an accumulation point that solves the original problem, accumulation will be caused by at least those cuts (during the solution process), that contribute to the determination of the approximate solution. The associated iterates correspond to a convergent subsequence $\{J_k , k=0,1,2,\ldots\}$ and have to be 'sufficiently dense' in $\{J=0,1,2,\ldots\}$. Higle and Sen succeeded in finding a formula for the expression 'sufficiently dense' and proved in [56] that in the case that

$$\lim_{k\to\infty} (J_{k-1}/J_k) = 1 \tag{27.3}$$

the corresponding accumulation point solves the problem.

In a next step the authors focus on a modification of the basis method (#0-#5) so to enforce property (27.3) and hence that any accumulation point solves the problem. The essential idea is to keep an iterate x^J and the corresponding cut alive as long as the iterate obtained by the, say, t-th iteration later, is not identified as 'clearly preferable' ('clearly preferable' is suggested to be controlled via suitable parameters). Those iterates that are kept alive during a fixed number of iterations are called *incumbent solutions* and the associated cuts *incumbent cuts*. These incumbent cuts will be fully updated periodically, i.e. with respect to the incumbent solution and with respect to the total sample information. In this way these incumbent cuts refer to the latest iteration. This fact obviously avoids that the incumbent cuts become negligible. Instead, these cuts are promoted for the next iterates as they are seen as essential for coming close to the desired solution. As a consequence all accumulation points of the sequence of incumbent solutions are solutions for the given problem. The resulting method is called *Stochastic Decomposition* (Higle and Sen 1991 [56]).

Similarly to SQG methods, stochastic decomposition might be stopped through statistical criteria that allow to verify optimality with respect to a prescribed tolerance and with respect to a probabilistic level. The authors suggest to focus on the progress of the incumbent objective values, on the values of θ that converge to 0 (see [56]) or on relying on subgradient estimates (Higle and Sen 1991 [57], [58]). In an advanced work Higle, Sen and Yakowitz 1990 [59] restrict the master problem to finitely many cuts, and obtain asymptotic optimality through incorporating a regularizing term.

As mentioned in the introduction of this work, nowadays, methods are available, which solve smooth, convex mathematical programs rather satisfactorily. So, it is close to investigate whether these methods might also be applied successfully to stochastic programs. Such algorithms usually require that the involved functions and their higher order information (gradient, Hessian) are calculable. This is a very strong assumption in the case that multidimensional integration is a major part of each iteration. As a consequence we have to distinguish cases in which

$$(E\rho)(x) = \int_{\Omega} \rho(\omega,x)dP(\omega) \tag{28.1}$$

and, in particular, its first order information is available. This is the case either for P being discrete or for highly structured stochastic programs with P being absolutely continuous.

In the case that P is discrete with a reasonable amount of atoms, then we already know that we may solve directly the deterministic equivalent problem; obviously, this requires algorithms available within mathematical programming. For the requirements posed on these methods we refer to section 7 .

For P being absolutely continuous and representable through a distribution function denoted $F(\omega)$, we recall investigations in Nazareth and Wets 1986 [102]. Slightly simplified to [102] we below consider stochastic linear programs with complete recourse and nonstochastic tenders $\chi=Tx$, i.e. the recourse function is of the form :

$$\rho(\omega,\chi) = \min \{q(\omega,y) \mid Wy= \omega-\chi \text{ , } y\geq 0\} \quad . \tag{28.2}$$

$y \rightarrow q(\omega,y)$ is supposed to be a real-valued convex (and hence continuous) function on $\mathbf{R}^{n'}$. This yields the deterministic equivalent problem

$$\min \langle c,x\rangle + (E\rho)(\chi) \tag{28.3.1}$$

$$\text{s.t.} \quad Ax = a \tag{28.3.2}$$

$$Tx - \chi = 0 \tag{28.3.3}$$

$$x\geq 0 \quad .$$

$Tx-\chi=0$ entails a decomposition of the objective into a linear and nonlinear term. The number of variables corresponding to the nonlinear term is determined by m' (the number of technological rows) in (28.3) being usually small compared with n (the number of components subject to which the objective is linear).

In the case of simple recourse (i.e. $W=[I,-I]$) let the recourse function of (28.2) be written as Lebesgue-Stieltjes integral :

$$(E\rho)(\chi) = \int_{\Omega} q(\omega,\omega-\chi)dF(\omega) \quad ; \tag{28.4}$$

due to $y \rightarrow q(\omega,y)$ being convex finite-valued on $\mathbf{R}^{n'}$ the subdifferential of $(E\rho)(\chi)$ is given as

$$\partial(E\rho)(\chi) = - \int_{\Omega} \partial q(\omega,\omega-\chi)dF(\omega) \tag{28.5}$$

with $\partial q(\omega,\omega-\chi)$ denoting the subdifferential with respect to the second argument (Nazareth and Wets 1986 [102]). If, additionally, the recourse costs in (28.2) are separable and independent of ω this entails separability of $(E\rho)(\chi)$ and hence the integral in (28.4) reduces to m' one-dimensional integrals

$$(E\rho)(\chi) = \Sigma_i \ (E\rho_i)(\chi_i) = \Sigma_i \int_{\Omega} q_i(\omega_i-\chi_i)dF_i(\omega) \tag{28.6}$$

where F_i denotes the marginal distribution function of ω_i ($i=1,\dots,m'$). The subdifferential of the convex function $E\rho$ is then the Cartesian product of the subdifferentials of $E\rho_i$ being given with

$$\partial(E\rho_i)(\chi_i) = - \int_{\Omega} \partial q_i(\omega_i-\chi_i)dF_i(\omega) \ . \tag{28.7}$$

Differentiability (smoothness) of $E\rho$ is obtained either with differentiability of $q(\omega,.)$, $q(.)$ respectively, or with absolute continuity of the random elements $q(\omega,.)$, ω respectively. Moreover, $E\rho$, $E\rho_i$ are convex, real-valued and thus continuous (Nazareth and Wets 1986 [102, p.5]).

For the case of separable, sublinear costs $q(y)$ being independent of ω and given for $i=1,\dots,m'$ according to

$$q_i(y_i) = \begin{cases} -q_i^- y_i & \text{for } y_i<0 \\ q_i^+ y_i & \text{for } y_i\geq 0 \end{cases} \tag{28.8}$$

Nazareth and Wets state expressions for $(E\rho_i)(\chi_i)$ and its subdifferential $\partial(E\rho_i)(\chi_i)$ with setting $q=q_i^-+q_i^+\geq 0$, $F_i^-(\chi_i)=F_i[\omega_i < \chi_i]$, $F_i(\chi_i)=F_i[\omega_i \leq \chi_i]$:

$$(E\rho_i)(\chi_i) = q_i^+. E\omega_i + (q_i F_i^-(\chi_i) - q_i^+).\chi_i - q_i \int_{\omega_i<\chi_i} \omega_i dF_i = \tag{28.9.1}$$

$$= q_i^+. E\omega_i + (q_i F_i(\chi_i) - q_i^+).\chi_i - q_i \int_{\omega_i\leq\chi_i} \omega_i dF_i \tag{28.9.2}$$

$$\partial(E\rho_i)(\chi_i) = [q_i F_i^-(\chi_i) - q_i^+ , q_i .F_i(\chi_i) - q_i^+] \quad . \tag{28.10}$$

In the case that $E\rho$ is differentiable Nazareth and Wets 1986 [102] (see also Nazareth and Wets 1988 [103]) recommend the application of smooth optimization methods for stochastic programs (28.3): in particular, the authors discuss algorithms based on the *generalized programming technique* according to Dantzig 1963 [24] and the *reduced gradient method* (Murtagh and Saunders 1978 [99]). For implementation issues on mathematical programming techniques we refer to Nazareth 1988 [101].

29 Scenarios and Policy Aggregation due to Rockafellar and Wets

The concept of *scenarios and policy aggregation* has been introduced for solving stochastic multi-stage problems in Rockafellar and Wets 1991 [127] and is strongly motivated to meet *'practitioners way '* of finding a good decision to the problem at hand: collecting some (finitely many) scenarios (i.e. possible outcomes of ω in future time periods) and assigning probabilities to these scenarios (representing how likely each scenario will appear), practitioners determine the optimal policy for each scenario and then intend to find *one* decision that hedges best (in a certain sense) against all possible scenarios relative to their probabilities.

In view of the fact that the present approach has been originally designed for stochastic multi-stage problems and in view of Remark 2.2 and its preceding interpretation of stochastic two-stage problems, we consider below two-stage problems within the multi-stage setting : this reveals the possibility to decide with respect to 'more (than no) information' in the first stage and to decide with respect to 'less (than full) information' in the second stage. Compared to the two-stage problems considered in the previous sections, this induces some relaxation on the first-stage decision and some restrictions on the second-stage decision.

Let P be discrete with Ω as finite support whose elements ω^s (s=1,...,N) are called *scenarios*; probabilities assigned to scenario ω^s are denoted $p(\omega^s)$; again x stands for the first-stage decision and y for the second-stage decision. We start with a stochastic two-stage program (in the multi-stage setting) of the form :

$$\min \ \Sigma_{\omega^s \in \Omega} \phi(\omega^s, x(\omega^s), y(\omega^s)).p(\omega^s) \ . \tag{29.1}$$

Now, decisions $x(.) \in \mathbf{R}^n$,$y(.) \in \mathbf{R}^{n'}$ are supposed to be taken with respect to a certain information being available in the first stage, in the second stage, respectively:

let the information in the first stage be given as partition $\{\Omega^{1,1},...,\Omega^{1,r'}\}$ of Ω; similarly, let the information in the second stage be given as partition $\{\Omega^{2,1},...,\Omega^{2,r''}\}$ of Ω. Up to the first stage we may have enough information to know in which $\Omega^{1,r}$ (for some $1 \leq r \leq r'$) the outcome will fall; however, at the first stage we shall not be able to distinguish the scenarios from each other within $\Omega^{1,r}$. Similarly, up to the second stage we may have (usually some more) information in which $\Omega^{2,r}$ (for some $1 \leq r \leq r''$) the outcome will fall; again, even in this second stage we shall not be able to distinguish the scenarios from each other within $\Omega^{2,r}$. Thus, as scenarios within $\Omega^{1,r}$ ($1 \leq r \leq r'$) are indistinguishable, these scenarios are (logically) required to share a common first-stage decision; analogously, scenarios within $\Omega^{2,r}$ ($1 \leq r \leq r''$) are required to share a common second-stage decision.

This requirement is called *nonanticipativity* and greatly complicates stochastic problems within this setting. Rockafellar and Wets called $\Omega^{1,r}$ ($1 \leq r \leq r'$), $\Omega^{2,r}$ ($1 \leq r \leq r''$) respectively, *scenario bundles* of the first stage, of the second stage respectively. The partitions $\{\Omega^{1,1},...,\Omega^{1,r'}\}$ and $\{\Omega^{2,1},..., \Omega^{2,r''}\}$ induce σ-algebras in both stages being denoted $\mathcal{W}^1$, $\mathcal{W}^2$. Naturally, supposing that information increases in time and will never be lost, scenario bundling is performed so that

$\mathcal{W}^1 \subset \mathcal{W}^2$, i.e. $\mathcal{W}^2$ is a refinement of $\mathcal{W}^1$. Mathematically nonanticipativity may be viewed as measurability condition for x(.), y(.) with respect to $(\Omega,\mathcal{W}^1)$, $(\Omega,\mathcal{W}^2)$.

Remark 29.1: Sofar, in stochastic two-stage programs we supposed to have *no information* about the outcome in the first stage, and *full information* of the outcome ω in the second stage. In terms of σ-fields this means that in the first stage we have the trivial σ-algebra $\mathcal{W}^1$ consisting only of two elements (the entire support Ω and the empty set $\emptyset$); in the second stage we have the finest σ-algebra $\mathcal{W}^2$ (being generated by the sets $\{\omega^s\}$, s=1,...,N).
Additionally, to nonanticipativity, we suppose to have for each scenario some feasibility set $Z(\omega^s)$ in $\mathbf{R}^{n+n'}$ subject to which decisions $x(\omega^s)$, $y(\omega^s)$ have to be chosen (as in [127]).

In accordance with Rockafellar and Wets let us denote $z(\omega^s)=(x(\omega^s),y(\omega^s))$ as *policy* corresponding to scenario ω^s. A policy may be seen as a mapping $z:\Omega\rightarrow\mathbf{R}^{n+n'}$ that assigns to each scenario ω^s a decision vector $(x(\omega^s),y(\omega^s))$. Let the associated vector space be denoted with $\mathcal{E}$. All those policies (x,y) that additionally satisfy $(x(\omega^s),y(\omega^s))\in Z(\omega^s)$ are denoted *admissible policies* and are collected in the set

$$\mathcal{A} := \{z\in\mathcal{E} \mid (x(\omega^s),y(\omega^s))\in Z(\omega^s) \text{ for all scenarios } \omega^s \} . \qquad (29.2)$$

Admissible policies need not be measurable w.r.t. $\mathcal{W}^1$, $\mathcal{W}^2$. The $(\mathcal{W}^1,\mathcal{W}^2)$-measurable elements of $\mathcal{E}$ are called *implementable policies* and form a subspace of $\mathcal{E}$ being defined according to [127] :

$$\begin{aligned}\mathcal{I} := \{z\in\mathcal{E} \mid\ & z=(x,y) \text{ with x constant on each bundle } \Omega^{1,r}\ (1\leq r\leq r'),\\ & \text{and y constant on each bundle } \Omega^{2,r}\ (1\leq r\leq r'')\} . \end{aligned}\qquad (29.3)$$

As a consequence of the above, one has to add to (29.1) those constraints that ensure that the optimal policy will be both admissible <u>and</u> implementable; this yields:

$$\begin{aligned}&\min\ \Sigma_{\omega^s\in\Omega}\, \phi(\omega^s,z).p(\omega^s)\\ &\text{s.t.} \qquad z\in\mathcal{A}\cap\mathcal{I} .\end{aligned}\qquad (29.4)$$

Having an arbitrary policy, say z=(x,y), one gets with

$$x(\Omega^{1,r}) := [\ \Sigma_{\omega^s\in\Omega^{1,r}}\ p(\omega^s).x(\omega^s)] \,/\, [\Sigma_{\omega^s\in\Omega^{1,r}}\, p(\omega^s)\ ]\ \ 1\leq r\leq r' , \qquad (29.5.1)$$

$$y(\Omega^{2,r}) := [\ \Sigma_{\omega^s\in\Omega^{2,r}}\ p(\omega^s).y(\omega^s)] \,/\, [\Sigma_{\omega^s\in\Omega^{2,r}}\, p(\omega^s)\ ]\ \ 1\leq r\leq r'' , \qquad (29.5.2)$$

weighted averages of all responses $x(\omega^s)$, $y(\omega^s)$ for scenarios in bundles $\Omega^{1,r}$ $(1 \leq r \leq r')$, $\Omega^{2,r}$ $(1 \leq r \leq r'')$. Performing further

$$\forall\ r: 1 \leq r \leq r' \quad x^{\mathcal{J}}(\omega^s) := x(\Omega^{1,r}) \quad \text{for all } \omega^s \in \Omega^{1,r} \tag{29.6.1}$$
$$\forall\ r: 1 \leq r \leq r'' \quad y^{\mathcal{J}}(\omega^s) := y(\Omega^{2,r}) \quad \text{for all } \omega^s \in \Omega^{2,r} \tag{29.6.1}$$

yields an implementable policy $z^{\mathcal{J}}(\omega^s) = (x^{\mathcal{J}}(\omega^s), y^{\mathcal{J}}(\omega^s)) \in \mathcal{J}$. The transformation

$$\Pi^{\mathcal{J}} := z \rightarrow z^{\mathcal{J}} \tag{29.7}$$

is a projection from $\mathcal{E}$ onto $\mathcal{J}$ (linear and satisfying $[\Pi^{\mathcal{J}}]^2 = \Pi^{\mathcal{J}}$), depending on the probabilities assigned to the scenarios ω^s. $\Pi^{\mathcal{J}}$ is called the *aggregation operator* relative to the given information structure and probabilities. Defining the inner product and norm within the vector space $\mathcal{E}$ as

$$<z,z'> = \Sigma_{\omega^s \in \Omega}\, z(\omega^s).z'(\omega^s).p(\omega^s) \ , \quad \|z\| = [\Sigma_{\omega^s \in \Omega} [z(\omega^s)]^2.p(\omega^s)]^{1/2} \tag{29.8}$$

entails $\Pi^{\mathcal{J}}$ to be an orthogonal projection onto the subspace $\mathcal{J}$. As a consequence the operator

$$\Pi^{\mathcal{M}} = I - \Pi^{\mathcal{J}} \tag{29.9}$$

is the orthogonal projection onto the subspace of $\mathcal{E}$ complementary to $\mathcal{J}$, denoted $\mathcal{M}$ ('I' represents the identity map). Using $\mathcal{J} = \{ z \in \mathcal{E} \mid \Pi^{\mathcal{M}} z = 0 \}$ in (29.4) yields

$$\begin{aligned} &\min\ \Sigma_{\omega^s \in \Omega}\, \phi(\omega^s, z).p(\omega^s) \\ &\text{s.t.} \quad z \in \mathcal{A}\, ,\ \Pi^{\mathcal{M}} z = 0\ . \end{aligned} \tag{29.10}$$

This problem (29.10) is not decomposable and likely becomes prohibitively large with the number of decision variables for each particular scenario. Due to this fact Rockafellar and Wets 1991 [127] recommend the socalled *Progressive Hedging Algorithm* for solving (29.10): this method is based on solving modified versions of the (considerably smaller) scenario problems for all ω^s

$$\min_{(x,y) \in Z(\omega^s)} \phi(\omega^s, x, y) \tag{29.11}$$

and aggregate the obtained solution according to (29.6) to an approximate solution for (29.10). Apparently, this method performs decentralized minimization, entailing that the complexity of solving problem (29.10) is reduced to the complexity of solving individual scenario problems.

In particular the method is stated in accordance with [127] :

Progressive Hedging Algorithm: In iteration J (J=0,1,2...) one has an admissible but not necessarily implementable policy $z^J \in \mathcal{A}$ and a price system $\zeta^{\mathcal{M},J} \in \mathcal{M}$. (Initially one can take z^0 to be the policy obtained by letting $z^0(\omega^s)$ be for each scenario an optimal solution to the given scenario problem (29.2); one can take $\zeta^{\mathcal{M},J}=0$.)

#1) Calculate the policy $z^{\mathcal{I},J} = \Pi^{\mathcal{I}} z^J$, which is implementable (but not necessarily admissible). If ever one wishes to stop, this policy $z^{\mathcal{I},J}$ is to be offered as the best substitute yet available for a solution to (29.11) .

#2) Calculate as z^{J+1} an (approximately) optimal solution where $z^{J+1}(\omega^s)$ for each scenario solves the subproblems

$$\min \ \phi(\omega^s,z) + z.\zeta^{\mathcal{M},J}(\omega^s) + (1/2)\gamma \, \| z - z^{\mathcal{I},J}(\omega^s) \|^2 \quad \text{over all } z \in Z(\omega^s) . \qquad (29.12)$$

The policy z^{J+1} will be admissible (but not necessarily implementable).

#3) Update from $\zeta^{\mathcal{M},J}$ to $\zeta^{\mathcal{M},J+1}$ by the rule $\zeta^{\mathcal{M},J+1}=\zeta^{\mathcal{M},J}+\gamma.\Pi^{\mathcal{M}} z^{J+1}$; the price system $\zeta^{\mathcal{M},J+1}$ again will be element of the subspace $\mathcal{M}$. Return to #1 with J:=J+1 .

Performing #1 is obvious according to the definition in (29.5)-(29.7). In #2 subproblems (29.12) result from the Lagrangian representation of (29.4) - via (29.10) - given by

$$\begin{aligned} &\min \ \Sigma_{\omega^s \in \Omega} \phi(\omega^s,z).p(\omega^s) + \langle \Pi^{\mathcal{M}} z , \zeta \rangle \\ &\text{s.t} \quad z \in \mathcal{A} , \zeta \in \mathcal{E} \quad . \end{aligned} \qquad (29.13)$$

Exploiting orthogonality entails $\langle \Pi^{\mathcal{M}}z,\zeta\rangle=\langle \Pi^{\mathcal{M}}z,\Pi^{\mathcal{M}}\zeta\rangle=\langle z,\Pi^{\mathcal{M}}\zeta\rangle$. These relations justify to concentrate on $\zeta^{\mathcal{M}}:=\Pi^{\mathcal{M}}\zeta$ instead of the entire price vector ζ. This price vector may be viewed as an estimate for the optimal (equilibrium) price system associated with nonanticipativity. Moreover, perturbations outside $\mathcal{M}$ may be taken care of through the selection of z as an implementable policy ([127], p.136). The authors therefore concentrate on perturbations within $\mathcal{M}$, define as Lagrangian

$$L(z,\zeta^{\mathcal{M}}) = \Sigma_{\omega^s \in \Omega} \phi(\omega^s,z).p(\omega^s) + \langle z,\zeta^{\mathcal{M}} \rangle \quad \text{for } z \in \mathcal{A}, \ \zeta^{\mathcal{M}} \in \mathcal{M} \qquad (29.14)$$

and pass then over to the *augmented* Lagrangian for $z \in \mathcal{A}$, $\zeta^{\mathcal{m}} \in \mathcal{M}$ being given through:

$$L_\gamma(z,\zeta^{\mathcal{m}}) = \Sigma_{\omega^s \in \Omega}\, \phi(\omega^s,z).p(\omega^s) + \langle z,\zeta^{\mathcal{m}}\rangle + (1/2)\, \gamma.\|\, \Pi^{\mathcal{m}} z \,\|^2 = \qquad (29.15)$$
$$= \Sigma_{\omega^s \in \Omega}\, \phi(\omega^s,z).p(\omega^s) + \langle z,\zeta^{\mathcal{m}}\rangle + (1/2)\, \gamma.\|\, z- z^{J} \,\|^2 .$$

Due to the fact that the term $\|\Pi^{\mathcal{m}} z\|^2$ is not decomposable into separate terms for each scenario, Rockafellar and Wets substitute $\|\Pi^{\mathcal{m}} z\|^2$ by $\|\Pi^{\mathcal{m}} z^J\|^2 = \|z- z^{J,J}\|^2$ that entails

$$\min\ \Sigma_{\omega^s \in \Omega}\, \phi(\omega^s,z).p(\omega^s) + \langle z,\zeta^{\mathcal{m}}\rangle + (1/2)\, \gamma.\|\, z- z^{J,J} \,\|^2 \qquad (29.16)$$

being decomposable into the scenario subproblems (29.12).

Due to the above, the authors pay attention to those policies (within the set of admissible policies) that solve a perturbed version of scenario subproblems; these policies are denoted *contingent policies*. For the mathematical details that ensure (29.5) to be solved with #1-#3 we refer to Rockafellar and Wets 1991 [127] and Robinson 1991 [112].

Remark 29.2: Wets 1989 [145] proved that the solutions' average (w.r.t. the probability weights) of #2 minimizes the epigraphical sum of the objectives in #2 (taken w.r.t. the probability weights).

The motivation to focus on the augmented Lagrangian instead on the classical Lagrangian stems from the algorithmic investigations (see Rockafellar 1976 [120]) within the area of mathematical programming. Therein, working with augmented Lagrangians has been proven to provide greater stability to solution methods, see Bertsekas 1982 [8]). The update rule in #3 for the projected price vector stems from the augmented Lagrangian theory and is part of the socalled multiplier method of first order. For its geometric interpretation and convergence analysis consult Bertsekas 1982 [8, p.104]). Further, in Rockafellar and Wets 1991 [127] it is stressed that the scenario subproblems (29.12) - with respect to any fixed scenario - within the sequence of iterations differ only in a linear term being part of the objectives and providing a simple way of updating (29.12) from one iteration to the next. Therefore employing parametric techniques likely entail a strong reduction in the computational effort.

A characteristic of the *progressive hedging algorithm* is that it produces in each iteration an implementable and an admissible solution, $z^{\hat{J},J}$ and z^J. The distance expression $\|z^J - z^{\hat{J},J}\|$ is taken as 'measure of how far one is from satisfying all the constraints' [127, p.126]. Under the assumptions that for each scenario ω^s the feasible set $Z(\omega^s)$ in $\mathbf{R}^{n+n'}$ is nonempty, convex and closed, and the objective function $\phi(\omega^s,.,.)$ is convex, locally Lipschitz continuous, inf-compact, the convergence theorems in [127] state : $\|z^J - z^{\hat{J},J}\|$ tend to 0 and at the same time the projected price vector will tend to an optimal solution of the Largrangian dual to problem (29.4). For the (restricted) linear-quadratic case (i.e. $\phi(\omega^s,.,.)$ being quadratic convex and $Z(\omega^s)$ being polyhedral convex) the convergence rate is proven to be linear (Theorem 5.2 in [127]). This convergence rate even holds in the case that in #2 the scenario subproblems are solved approximately based on a criterion, indicating whether 0 is sufficiently close to the subdifferential (of the objective) at $z^J(\omega^s)$ with respect to the distance $\| z^J(\omega^s) - z^{\hat{J},J}(\omega^s) \|$ (Rockafellar and Wets 1991 [127, Theorem 5.4]).

Extensions and improvements of scenarios and policy aggregation are investigated in Robinson 1991 [112], and Rockafellar and Wets 1990 [126]: In [112] nonseparable convex constraints are incorporated in (29.4); [126] aims at getting better estimates for the dual solutions associated with the nonanticipativity.

For implementation issues on the progressive hedging algorithm and its application to stochastic multi-stage networks we refer to Mulvey and Vladimirou 1989 [97], 1991 [98].

V *BRAIN* - *BaRycentric Approximation for INtegrands (Implementation Issues)*

In this chapter we present **BRAIN**, an implementation (written in Fortran77) of the **BaRycentric** Approximation for **IN**tegrands. In chapter III we mainly focused on understanding and properties of the approximation. Below, we concentrate to ease its *usage* and to illustrate the *effort* of the computational activities within one iteration of the algorithm stated in section 20. Moreover, that which follows gives also some insight in how to exploit the structural properties of a $\times$-simplex.

We will restrict ourselves to the case of $\Theta\times\Xi$ representing a Cartesian product of (finitely many) closed, regular simplices compatible with the stochastic independence of random data. Therefore the case of Ω being a rectangle is covered only if all components of ξ, of η respectively, are stochastically independent. Further, we suppose that all approximate problems in algorithm (20.1) are solvable, primally and dually. (It only makes sense to exchange data if these exist.)

In accordance with chapter III the stochastic program we refer to (throughout this chapter) is written as

$$\inf_{x\in X} \quad f(x) + \int_{\times_i (\Theta^i\times\Xi^i)} \rho(\times_i (\eta^i,\xi^i),x)\, d(\times_i P^i(\eta^i,\xi^i)) \quad . \tag{V.1}$$

Remark V.1: Note that in case X is a singleton, i.e. X consists of exactly one element, the optimization problem V.1 reduces to the integration of a saddle function. This fact strongly broadens the applicability of BRAIN.

We start (section 30) with outlining the data structure for an arbitrary *probability measure* P of ω that is characterizable through a finite number of parameters. We exploit stochastic independences among ω^i (i=1,...,I) within evaluations and storages of their *(initial) extremal marginal distributions* (section 31). We then pass over to describe the data structure for the solutions of the *(initial) outer* and *inner approximation* (section 32) serving as interface to an optimization package in the case of solving (V.1) . Subsequently (section 33), we define data for the $\times$-*simplicial partition* and report on the extensions of data structures corresponding to *extremal distributions* (section 34), corresponding to the solutions of *outer* and *inner approximation* respectively (section 35). We conclude this chapter with outlining the main ideas for implementing the $\times$-*simplicial refinement* (section 36).

Remark V.2: The reason for distinguishing between '(initial) extremal marginal distribution' and the 'extremal distribution' in subsequent iterations is simply to stress and to ease the understanding that the latter is derived from the conditional probability measure of P^i with respect to a sub-$\times$-simplex $\Theta^{i,t} \times \Xi^{i,t'}$ and not with respect to the entire support $\Theta^i \times \Xi^i$ of $\omega^i=(\eta^i,\xi^i)$; as a consequence this extremal distribution is no longer 'marginal', like the 'initial' one. This fact has become already evident in (16.10).

Remark V.3: The proposed data structures have been designed with the intention to set up and cover the essential information required for the usage of the barycentric approximation scheme. In particular we set up after the transformation outlined in section 4, that creates the saddle property of the recourse function. Therefore, application to other approaches necessitates appropriate adaptions. Further, the data structures will be partitioned into *header* and *body*, whereby the 'header' stores characteristic indicators enabling to identify in which way the 'header' itself and the 'body' ought to be read.

As the type of probability measures P^i, the type of distribution functions respectively, given within real problems might vary for i=1,...,I, we have to build up the associated data structure for their 'input', denoted **distr.inp**, rather generally:

Distr.inp stores all marginal distributions given with P^i (i=1,...,I) and implicitly defining P due to their stochastic independences. The different types of distributions are controlled by means of *nswtyp(i)*, representing a certain kind of switch identifying the 'type' of P^i based on the following way of coding:

nswtyp(i)=0 iff P^i is a sample with treatable finite number of realizations; the corresponding probability is given implicitly through sample size N^i.

nswtyp(i)=1 iff P^i is a discrete distribution with treatable finite number of atoms; the associated probabilities have to be given explicitly.

nswtyp(i)=2 iff P^i is Normal distributed.

nswtyp(i)=3 iff P^i is Log-Normal distributed.

nswtyp(i)=4 iff P^i is Beta distributed.

nswtyp(i)=5 iff P^i is Weibull distributed.

(etc.)

Further classical distributions are to be treated analogously.

Depending on the type of the distribution we have to store in the *body* either the realizations of the sample, or the atoms of the discrete distribution together with the probability, or simply the parameters defining the given 'classical' distributions.

Name : DISTR.INP

As *Header* we specify :

I	: number of subvectors being stochastically independent ;
nswtyp(i)	: switch identifying the type of distribution P^i , i=1,...,I ;
K_i , L_i	: dimensions of the random-vectors η^i,ξ^i , i=1,...,I ;
N^i	: sample size of ω^i, in the case *nswtyp(i)*=0, number of atoms of P^i , in the case *nswtyp(i)*=1, number of parameters defining P^i , in the case *nswtyp(i)*≥2 , (i=1,...,I) ;
a_{ν_i} , b_{μ_i}	: vectors of the vertices defining the simplices Θ^i, Ξ^i covering the support of η^i,ξ^i sufficiently accurate ($\nu_i = 0,...,K_i$, $\mu_i = 0,...,L_i$, i=1,...,I) .

In the *Body* we store sequentially for i=1,...,I:

if *nswtyp(i)*=0:

$(\eta^i,\xi^i)^r$	: vector of realizations stored sequentially, r=1,...,N^i , ((K_i+L_i) values per realization r and per subvector i);

if *nswtyp(i)*=1:

$(p_i ,(\eta^i,\xi^i))^r$	: probabilities, vector of atoms stored sequentially, r=1,...,N^i , (1+(K_i+L_i) values per atom r and per subvector i);

if *nswtyp(i)*=2,3...:

par_i	: N^i-dimensional vector of parameters, stored sequentially, defining the 'classical' distributions (N^i values per subvector i).

ENDDATA (Distr.inp)

Remark 30.1: The simplices in the header of *distr.inp* have to be defined in accordance with uniform integrability of section 3, stating that realizations outside a certain bounded region have negligible influence to the underlying problem. As quite often in practice, distributions with unbounded support (e.g. Normal distributions) provide sufficiently well approximates for 'real' distributions with bounded support, we are rather convinced that appropriate simplices might be specified with an affordable effort individually, focusing on the 'real problem' at hand.

31 *Evaluation of Initial Extremal Marginal Distributions*

Employing the notation and formulae (16.8)-(16.11) of section 16 for the initial iteration (J=0) we have for i=1,...,I

$J=0;\ t^J=t^0=1,$

$P_{l,i,t,J} = P_{l,i,1,0}$ (initial lower extremal marginal distribution)

$P_{u,i,t,J} = P_{u,i,1,0}$ (initial upper extremal marginal distribution)

$P_{l,t,J} = P_{l,1,0} = P_{l,0}$ (initial lower extremal distribution)

$P_{u,t,J} = P_{u,1,0} = P_{u,0}$ (initial upper extremal distribution)

$$a_{\nu_i}^{r(i,t,J)} = a_{\nu_i}^{r(i,1,0)} =: a_{\nu_i}^{0} \qquad \nu_i = 0,\dots,K_i \qquad (31.1.1)$$

$$b_{\mu_i}^{s(i,t,J)} = b_{\mu_i}^{s(i,1,0)} =: b_{\mu_i}^{0} \qquad \mu_i = 0,\dots,L_i \qquad (31.1.2)$$

$$\xi_{\nu_i}^{s(i,t,J)} = \xi_{\nu_i}^{s(i,1,0)} =: \xi_{\nu_i}^{0} \qquad \nu_i = 0,\dots,K_i \qquad (31.1.3)$$

$$\eta_{\mu_i}^{r(i,t,J)} = \eta_{\mu_i}^{r(i,1,0)} =: \eta_{\mu_i}^{0} \qquad \mu_i = 0,\dots,L_i \qquad (31.1.4)$$

$$p_{\Theta}^{(i,t,J)}(\nu_i) = p_{\Theta}^{(i,1,0)}(\nu_i) =: p_{\Theta}^{0}(\nu_i) \qquad \nu_i = 0,\dots,K_i \qquad (31.1.5)$$

$$p_{\Xi}^{(i,t,J)}(\mu_i) = p_{\Xi}^{(i,1,0)}(\mu_i) =: p_{\Xi}^{0}(\mu_i) \qquad \mu_i = 0,\dots,L_i\,. \qquad (31.1.6)$$

Let us introduce the matrix of cross moments, denoted $Mt^{i,0}$, for i=1,...,I

$$Mt^{i,0,} := \begin{pmatrix} m_{0,0}^{i,0} & m_{0,1}^{i,0} & \dots & m_{0,L_i}^{i,0} \\ \cdot & \cdot & \cdot & \cdot \\ m_{K_i,0}^{i,0} & m_{K_i,1}^{i,0} & \dots & m_{K_i,L_i}^{i,0} \end{pmatrix} \qquad (31.2)$$

being defined for i=1,...,I according to

$$m_{0,0}^{i,0} = 1 \qquad (31.3.1)$$

$$m_{k,0}^{i,0} = \int_{\Theta^i \times \Xi^i} \eta_k^{\,i}\, dP^i \qquad k=1,\dots,K_i \qquad (31.3.2)$$

$$m_{0,l}^{i,0} = \int_{\Theta^i \times \Xi^i} \xi_l^i \, dP^i \qquad l=1,\dots,L_i \, , \tag{31.3.3}$$

$$m_{k,l}^{i,0} = \int_{\Theta^i \times \Xi^i} \eta_k^i \, \xi_l^i \, dP^i \quad k=1,\dots,K_i \, , \, l=1,\dots,L_i \quad . \tag{31.3.4}$$

We start with the evaluation of the extremal marginal distributions denoted $P_{l,i,1,0}$ and $P_{u,i,1,0}$ for P^i; herewith, $P_{l,i,1,0}$ and $P_{u,i,1,0}$ represent the marginal distributions for the desired $P_{l,0}$ and $P_{u,0}$. Due to the formulae (16.8)-(16.11) in section 16 we therefore have to compute for all i=1,...,I : the atoms of $P_{l,i,1,0}$, $P_{u,i,1,0}$ denoted $(a_{\nu_i}{}^0, \xi_{\nu_i}{}^0)$, $(\eta_{\mu_i}{}^0, b_{\mu_i}{}^0)$ and probabilities $p_\Theta{}^0(\nu_i)$, $p_\Xi{}^0(\mu_i)$ where $a_{\nu_i}{}^0$, $b_{\mu_i}{}^0$ represent the vertices of Θ^i, Ξ^i ($\nu_i=0,\dots,K_i$, $\mu_i=0,\dots,L_i$). This yields for i=1,...,I :

$$p_\Theta{}^0(\nu_i) = \sum_{k=0}^{K_i} [S^{-i}]_{\nu_i,k} . m_{k,0}^{i,0} \tag{31.4.1}$$

$$p_\Xi{}^0(\mu_i) = \sum_{l=0}^{L_i} [T^{-i}]_{\mu_i,l} . m_{0,l}^{i,0} \tag{31.4.2}$$

$$\xi_{\nu_i}{}^0 = \{ \sum_{\mu_i=0}^{L_i} b_{\mu_i}{}^0 \, [\sum_{k=0}^{K_i} \sum_{l=0}^{L_i} [S^{-i}]_{\nu_i,k} . [T^{-i}]_{\mu_i,l} \cdot m_{k,l}^{i,0} \;] \; \} . \{ p_\Theta{}^0(\nu_i) \}^{-1} \tag{31.4.3}$$

$$\eta_{\mu_i}{}^0 = \{ \sum_{\nu_i=0}^{K_i} a_{\nu_i}{}^0 \, [\sum_{k=0}^{K_i} \sum_{l=0}^{L_i} [S^{-i}]_{\nu_i,k} . [T^{-i}]_{\mu_i,l} \cdot m_{k,l}^{i,0} \;] \; \} . \{ p_\Xi{}^0(\mu_i) \}^{-1} \; . \tag{31.4.4}$$

S^{-i}, T^{-i} denote the inverses of the regular matrices S^i, T^i corresponding to the (regular) simplices Θ^i, Ξ^i. Thus, for getting the extremal marginal distributions $P_{l,i,1,0}$ and $P_{u,i,1,0}$ we have to invert S^i, T^i and evaluate the integrals in (31.3). Depending on the type of distributions P^i we either have to compute a finite sum (in the case that P^i is a sample , discrete with finite support respectively) or we have to employ efficient sampling techniques for **one-** and **two-dimensional integrals** (in the case that P^i is absolutely continuous or with support being too large with respect to the number of realizations).

Remark 31.1: In the case that some of the probabilities $p_\Xi{}^0(\mu_i)$, $p_\Theta{}^0(\nu_i)$ are 0 we may choose any point within the corresponding simplices Θ^i, Ξ^i as $\eta_{\mu_i}{}^0$, $\xi_{\nu_i}{}^0$ (see Theorems 14.3 and 14.3').

Due to formulae (31.4.1)-(31.4.4) it becomes evident that the numerical operations for determining probabilities and support can be kept low in the case of a tricky usage of S^{-i}, T^{-i} and of the cross moment matrix $Mt^{i,0}$ (i=1,...,I) .

$P_{l,0}$ and $P_{u,0}$ associated with P are achieved with

$$P_{l,0} = \mathop{X}_{i=1}^{I} P_{l,i,1,0} \qquad (31.5.1)$$

$$P_{u,0} = \mathop{X}_{i=1}^{I} P_{u,i,1,0} \ . \qquad (31.5.2)$$

(31.5) yields atoms (a_v^0, ξ_v^0) and associated probabilities $p_\Theta^0(v)$ for $P_{l,0}$ with $v=(v_1,...,v_I)$, $v_i=0,...,K_i$:

$$(a_v^0, \xi_v^0) = \mathop{X}_{i=1}^{I} (a_{v_i}^0, \xi_{v_i}^0) \qquad (31.6.1)$$

$$p_\Theta^0(v) = \prod_{i=1}^{I} p_\Theta^0(v_i) \ . \qquad (31.6.2)$$

and analogously, atoms (η_μ^0, b_μ^0) and associated probabilities $p_\Xi^0(\mu)$ for $P_{u,0}$ with $\mu=(\mu_1,...,\mu_I)$, $\mu_i=0,...,L_i$:

$$(\eta_\mu^0, b_\mu^0) = \mathop{X}_{i=1}^{I} (\eta_{\mu_i}^0, b_{\mu_i}^0) \qquad (31.7.1)$$

$$p_\Xi^0(\mu) = \prod_{i=1}^{I} p_\Xi^0(\mu_i) \ . \qquad (31.7.2)$$

The cardinality of the supports of $P_{l,0}$ and $P_{u,0}$ are therefore given with

$$\prod_{i=1}^{I} (K_i+1) \quad \text{and} \quad \prod_{i=1}^{I} (L_i+1) \ . \qquad (31.8)$$

We define next **ledd.dsc** (lower extremal discrete distribution) and **uedd.dsc** (upper extremal discrete distribution) for storing the extremal distributions $P_{l,0}$ and $P_{u,0}$:

Name: LEDD.DSC

In the *Header* we specify

I	: number of subvectors being stochastically independent ;
K_i , L_i	: dimensions of the random vectors η^i, ξ^i , i=1,...,I ;
nat	: total number of atoms (= $\Pi_i(K_i+1)$) - due to (31.8) .

In the *Body* we specify for each atom ν :

$p_\Theta{}^0(\nu)$	: probability ;
$(a_\nu{}^0, \xi_\nu{}^0)$	: Σ_i (K_i +L_i) components of atom.

ENDDATA (ledd.dsc)

Name: UEDD.DSC

In the *Header* we specify

I	: number of subvectors being stochastically independent ;
K_i , L_i	: dimensions of the random vectors η^i, ξ^i , i=1,...,I ;
nat	: total number of atoms (= $\Pi_i(L_i+1)$) - due to (31.8) .

In the *Body* we specify for each atom μ :

$p_\Xi{}^0(\mu)$	: probability ;
$(\eta_\mu{}^0, b_\mu{}^0)$	: Σ_i (K_i+L_i) components of atom .

ENDDATA (uedd.dsc)

Due to (20.1.3) let the lower extremal distribution $P_{l,0}$ substitute P in (V.1). Thus we obtain $(LAP)^0$ as Lower Approximating Problem in the initial iteration (J=0):

$$(LAP)^0 \qquad \inf_{x \in X} \; f(x) + \Sigma_\nu \, \rho(a_\nu^0, \xi_\nu^0, x).p_\Theta^0(\nu) \quad . \qquad (32.1)$$

In the case that (V.1) is a mathematical program, then $(LAP)^0$ is a blockstructured program. Due to this block structure $(LAP)^0$ ought to be solved with specially designed algorithms as already stated in section 7.

The obtained optimal solution, subgradients respectively, and other essential information being required within several steps of the procedure (20.1) will be stored in **outap.sol** (**out**er **ap**proximation **sol**ution):

Name: OUTAP.SOL

In the *Header* we specify

nat	: number of atoms ($= \Pi_i \; (K_i+1)$) - according to (31.8) ;
n	: dimension of the first stage decision x ;
$\Sigma_i K_i$	: dimension of $(\eta^1,...,\eta^I)$;
$\Sigma_i L_i$	: dimension of $(\xi^1,...,\xi^I)$;
$\Psi^{l,0}$	: optimal objective value of (32.1) ;
$x^{opt,0}$	: optimal solution (n components) of (32.1) .

In the *Body* we specify for each atom ν :

$p_\Theta^0(\nu)$	: probability of atom ;
$\rho(a_\nu^0,\xi_\nu^0,x^{opt,0})$	: second-stage value ;
$\eta^*(a_\nu^0,\xi_\nu^0,x^{opt,0})$	: subgradient of $\rho(\eta,\xi,x^{opt,0})$ w.r.t. η at (a_ν^0,ξ_ν^0) ;
$\xi^*(a_\nu^0,\xi_\nu^0,x^{opt,0})$	: subgradient of $\rho(\eta,\xi,x^{opt,0})$ w.r.t. ξ at (a_ν^0,ξ_ν^0) ;
(a_ν^0,ξ_ν^0)	: vector of atom .

ENDDATA (outap.sol)

An upper bound for the objective value at $x^{opt,0}$ of problem (V.1) is given with the Upper Approximating Problem $(UAP)^0$, obtained with substituting P by the upper extremal discrete distribution $P_{u,0}$ and evaluating :

$$(UAP)^0 \qquad f(x^{opt,0}) + \Sigma_\mu \, \rho(\eta_\mu{}^0, b_\mu{}^0, x^{opt,0}).p_\Xi{}^0(\mu) \quad . \qquad (32.2)$$

In the case that (V.1) is a stochastic program with recourse requiring minimization within the second stage, $(UAP)^0$ decomposes into $\Pi_i(L_i+1)$ small and independently solvable programs. Optimal solutions, subgradients and further information required for keeping the machinery running will be stored in **innap.sol (inner approximation solution)**.

Name: INNAP.SOL

In the *Header* we specify

nat	: number of atoms ($= \Pi_i (L_i+1)$) - according to (31.8) ;
n	: dimension of the first-stage decision 'x' ;
$\Sigma_i K_i$	: dimension of $(\eta^1,...,\eta^I)$;
$\Sigma_i L_i$	: dimension of $(\xi^1,...,\xi^I)$;
$\Psi^{u,0}$	: objective value of (32.2) at $x^{opt,0}$;
$x^{opt,0}$	: first-stage decision to which the upper bound refers.

In the *Body* we specify for each atom μ :

$p_\Xi{}^0(\mu)$	: probability of atom;
$\rho(\eta_\mu{}^0,b_\mu{}^0,x^{opt,0})$	: second-stage value;
$\eta^*(\eta_\mu{}^0,b_\mu{}^0,x^{opt,0})$	: subgradient of $\rho(\eta,\xi,x^{opt,0})$ w.r.t. η at $(\eta_\mu{}^0,b_\mu{}^0)$;
$\xi^*(\eta_\mu{}^0,b_\mu{}^0,x^{opt,0})$	: subgradient of $\rho(\eta,\xi,x^{opt,0})$ w.r.t. ξ at $(\eta_\mu{}^0,b_\mu{}^0)$;
$(\eta_\mu{}^0,b_\mu{}^0)$	: vector of atom .

ENDDATA(innap.sol)

According to the refinement strategy of section 19 we investigate the 'degree' of nonlinearity of the integrand at $x^{opt,0}$ and hence store the support points $(a_v{}^0,\xi_v{}^0)$, $(\eta_\mu{}^0,b_\mu{}^0)$ and subgradients $\eta^*(a_v{}^0,\xi_v{}^0,x^{opt,0})$, $\xi^*(a_v{}^0,\xi_v{}^0,x^{opt,0})$, $\eta^*(\eta_\mu{}^0,b_\mu{}^0,x^{opt,0})$, $\xi^*(\eta_\mu{}^0,b_\mu{}^0,x^{opt,0})$ of the function $\rho(.,.,x^{opt,0})$. So, *outap.sol* will be responsible mainly for measuring the degree of nonlinearity in η, where *innap.sol* will provide information about nonlinearity of $\rho(.,.,x^{opt,0})$ in ξ .

Having decomposed ω into I subvectors to exploit stochastic independence we have to work with 2*I (closed and regular) simplices Θ^i, Ξ^i for $i=1,...,I$, containing the support of $\omega^i=(\eta^i,\xi^i)$. To improve the approximation we apply the ×-simplicial refinement scheme (16.14)-(16.16) in section 16. Herein we stress that partitioning some sub-×-simplex (closed, open or half-open) of $\Theta\times\Xi$ yields exactly two new sub-×-simplices (closed, open or half-open) through splitting exactly 1 edge of exactly one (of the 2*I) simplices. This refinement scheme raises a binary tree with the following characterization :

a) the root-×-simplex $\Theta\times\Xi=\times_i[\Theta^i\times\Xi^i]$ is characterized by the simplices Θ^i, Ξ^i ;
b) any node that is not a leaf of that tree is the 'father' of two 'sons';
c) each node represents exactly one sub-×-simplex ;
d) the union of all sub-×-simplices whose nodes represent leaves yields $\Theta\times\Xi$;
e) the leaves (i.e.: the associated sub-×-simplices) are mutually disjoint .

Remark 33.1: Whenever we speak of the 'root', of a 'node' or of a 'leaf', we actually think of the associated sub-×-simplex. We call the binary tree **partition tree.**

Remark 33.2: Having in mind the barycentric approximation scheme, it is evident that only leaves contribute to the lower and upper bounds of problem (V.1) through their extremal distributions.

Due to the features of the barycentric approximation scheme we suggest to create the following 'pots' for storing and handling the required information:

i) VERTEX POT
ii) SIMPLEX POT
iii) NODE POT
iv) LEAF POT

i) VERTEX POT

Vertex pot consists of '2*I' subpots being denoted VERTX'i'1.POT and VERTX'i'2 where i=1,...I points to Θ^i, (e.g. VERTX31.POT for i=3) and to Ξ^i (e.g. VERTX42.POT for i=4). These pots are of the same structure:

Name: VERTX'i'1.POT

In the *Header* we specify

K_i : dimension of Θ^i ;
$nvt1_i$: number of vertices in current nested simplicial collection of Θ^i.

In the *Body* we sequentially store the vertices $vt1=1,...,nvt1_i$:

a(vt1) : K_i components of the (vt1)-th vertex .

ENDDATA (vertx'i'1.pot)

Name: VERTX'i'2.POT

In the *Header* we specify

L_i : dimension of Ξ^i ;
$nvt2_i$: number of vertices in current nested simplicial collection of Ξ^i .

In the *Body* we sequentially store the vertices $vt2=1,...,nvt2_i$:

b(vt2) : the L_i components of the (vt2)-th vertex .

ENDDATA (vertx'i'2.pot)

ii) SIMPLEX POT

Similarly to the above, *simplex pot* consists of '2*I' subpots denoted SIMPX'i'1.POT and SIMPX'i'2.POT, where again i=1,...I points to Θ^i (e.g. SIMPX41 for i=4) and to Ξ^i (s.t. SIMPX32.POT for i=3). These pots have the same structure:

Name: SIMPX'i'1.POT

In the *Header* we specify

K_i : dimension of Θ^i ;
$nsx1_i$: number of subsimplices in current nested simplicial collection of Θ^i .

In the *Body* we sequentially store the subsimplices $sx1=1,...,nsx1_i$:

nrv1(sx1) : (K_i+1) integers stating the number of vertices - with respect to *vertx'i'1.pot* - that define the (sx1)-th subsimplex .

ENDDATA(simpx'i'1.pot)

Name: SIMPX'i'2.POT

In the *Header* we specify

L_i : dimension of Ξ^i ;
$nsx2_i$: number of subsimplices in current nested simplicial collection of Ξ^i .

In the *Body* we sequentially store the subsimplices $sx2=1,...,nsx2_i$:

nrv2(sx2) : (L_i+1) integers stating the number of vertices - with respect to *vertx'i'2.pot* - that define the (sx2)-th subsimplex .

ENDDATA (simpx'i'2.pot)

iii) NODE POT

Node pot characterizes the nodes (i.e. sub-×-simplices) in the *partition tree*:

Name: NODE.POT

In the *Header* we specify

I	: number of stochastically independent subvectors ;
nnode	: number of nodes (sub-×-simplices) in current partition tree.

In the *Body* we specify for any node nd=1,...,nnode :

ns1(nd)	: vector of I integers stating the numbers of subsimplices - s.t. *simpx'i'1.pot* - defining the sub-×-simplex within Θ ;
ns2(nd)	: vector of I integers stating the numbers of subsimplices - s.t. *simpx'i'2.pot* - defining the sub-×-simplex within Ξ ;
nswitch(nd)	: switch for partitioning Θ^i (nswitch=1) or Ξ^i (nswitch=2) (default 0 implies : this node is a leaf) ;
ptsx(nd)	: represents a pointer to the subsimplex to be partitioned (default 0 implies : this node is a leaf) ;
ptv1(nd),ptv2(nd)	: represents two pointers to those vertex numbers - of subsimplex ptsx(nd) - whose edge is to be partitioned (default 0,0 implies : this node is a leaf) ;
wt1(nd),wt2(nd)	: two 'weights' defining the new vertex of the partitioned edge (default : 0.5) ;
nv3(nd)	: the new vertex number (either in *vertx'i'1.pot* if nswitch(nd)=1 or in *vertx'i'2.pot* if nswitch(nd)= 2 ; 'i' is 'ptsx(nd)') (default 0 implies that this node is a leaf) ;
npt0(nd)	: pointer to 'father node' in the *partition tree* (0 indicates root);
npt1(nd)	: pointer to 'left son' within the *partition tree* (default 0 implies: this node is a leaf) ;
npt2(nd)	: pointer to 'right son' within the *partition tree* (default 0 implies: this node is a leaf) .

ENDDATA (node.pot)

Remark 33.4: In the case of highly structured problems (e.g. simple recourse) and one-dimensional simplices we suggest to choose for wt1(nd),wt2(nd) the barycentric weights of the corresponding conditional expectation; in this way we get the exact expectation of the integrand with respect to the underlying iterate instead of merely an upper bound (due to Kall and Stoyan 1982 [79]; see also section 21).

iv) LEAF POT

Leaf pot stores the pointers identifying leaves of *partition tree* in *node pot*:

Name: LEAF.POT

In the *Header* we specify

I	: number of stochastically independent subvectors ;
nnode	: number of nodes in current *partition tree* ;
nleaf	: number of leaves of the current *partition tree* ;
nsplit	: number of leaves (i.e. sub-×-simplices) to be split .

In the *Body* we specify sequentially for each leaf, t=1,...,nleaf :

npt(t)	: pointer of leaf t identifying its node in *node pot* ;
prob(t)	: probability of leaf 't' (i.e. sub-×-simplex) .

ENDDATA(leaf.pot)

Remark 33.5: In iteration J *leaf pot* yields nleaf=t^J in accordance with section 16.

In accordance with section 16 we enumerate with $t=1,\dots,t^J$ the leaves of the *partition tree*. The associated sub-×-simplices will be denoted $\Omega^{t,J}$ where for the ease of exposition

$$\Omega^{i,t,J} := (\Theta^{r(i,t,J)} \times \Xi^{s(i,t,J)}) \tag{34.1.1}$$

$$\Omega^{t,J} := \times_i (\Theta^{r(i,t,J)} \times \Xi^{s(i,t,J)}) . \tag{34.1.2}$$

Let the conditional distribution of P^i subject to $\Theta^{r(i,t,J)} \times \Xi^{s(i,t,J)}$ be denoted $P^{i,t,J}$. We then denote the moment matrices with respect to $\Theta^{r(i,t,J)} \times \Xi^{s(i,t,J)}$ $(i=1,\dots,I;\ t=1,\dots,t^J)$:

$$Mt^{i,t,J} := \begin{pmatrix} m_{0,0}^{i,t,J} & m_{0,1}^{i,t,J} & \dots & m_{0,L_i}^{i,t,J} \\ . & . & . & . \\ m_{K_i,0}^{i,t,J} & m_{K_i,1}^{i,t,J} & \dots & m_{K_i,L_i}^{i,t,J} \end{pmatrix} \tag{34.2}$$

where the coefficients are defined according to the cross moments

$$m_{0,0}^{i,t,J} = 1 \tag{34.3.1}$$

$$m_{k,0}^{i,t,J} = \int_{\Omega^{i,t,J}} \eta_k^i \, dP^{i,t,J} \qquad k=1,\dots,K_i \tag{34.3.2}$$

$$m_{0,l}^{i,t,J} = \int_{\Omega^{i,t,J}} \xi_l^i \, dP^{i,t,J} \qquad l=1,\dots,L_i \tag{34.3.3}$$

$$m_{k,l}^{i,t,J} = \int_{\Omega^{i,t,J}} \eta_k^i \xi_l^i \, dP^{i,t,J} \qquad k=1,\dots,K_i;\ l=1,\dots,L_i \ . \tag{34.3.4}$$

We continue with the evaluation of the extremal distributions denoted $P_{l,i,t,J}$, $P_{u,i,t,J}$ for $P^{i,t,J}$ subject to $\Omega^{i,t,J}$. Due to the formulae in section 16 we therefore have to compute for $i=1,\dots,I$: the atoms of $P_{l,i,t,J}$, $P_{u,i,t,J}$ denoted $(a_{\nu_i}^{r(i,t,J)}, \xi_{\nu_i}^{s(i,t,J)})$, $(\eta_{\mu_i}^{r(i,t,J)}, b_{\mu_i}^{s(i,t,J)})$ and the associated probabilities $p_\Theta^{r(i,t,J)}(\nu_i)$, $p_\Xi^{s(i,t,J)}(\mu_i)$ where $a_{\nu_i}^{r(i,t,J)}$, $b_{\mu_i}^{s(i,t,J)}$ are the vertices of $\Theta^{r(i,t,J)}$, $\Xi^{s(i,t,J)}$ $(\nu_i=0,\dots,K_i,\ \mu_i=0,\dots,L_i)$. We get

$$p_\Theta^{r(i,t,J)}(v_i) = \sum_{k=0}^{K_i} [S^{-r(i,t,J)}]_{v_i,k} . m_{k,0}^{i,t,J} \quad (34.4.1)$$

$$p_\Xi^{s(i,t,J)}(\mu_i) = \sum_{l=0}^{L_i} [T^{-s(i,t,J)}]_{\mu_i,l} . m_{0,l}^{i,t,J} \quad (34.4.2)$$

$$\xi_{v_i}^{s(i,t,J)} = \{ \sum_{\mu_i=0}^{L_i} b_{\mu_i}^{s(i,t,J)} [\sum_{k=0}^{K_i} \sum_{l=0}^{L_i} [S^{-r(i,t,J)}]_{v_i,k} . [T^{-s(i,t,J)}]_{\mu_i,l} . m_{k,l}^{i,t,J}] \} . \{p_\Theta^0(v_i)\}^{-1} \quad (34.4.3)$$

$$\eta_{\mu_i}^{r(i,t,J)} = \{ \sum_{v_i=0}^{K_i} a_{v_i}^{r(i,t,J)} [\sum_{k=0}^{K_i} \sum_{l=0}^{L_i} [S^{-r(i,t,J)}]_{v_i,k} . [T^{-s(i,t,J)}]_{\mu_i,l} . m_{k,l}^{i,t,J}] \} . \{p_\Xi^0(\mu_i)\}^{-1} \quad (34.4.4)$$

$S^{-r(i,t,J)}$, $T^{-S(i,t,J)}$ denote the inverses of the regular matrices $S^{r(i,t,J)}$, $T^{s(i,t,J)}$ associated with the (regular) simplices $\Theta^{r(i,t,J)}$, $\Xi^{s(i,t,J)}$. Now, analogously to the initial iteration, we invert $S^{r(i,t,J)}$,$T^{s(i,t,J)}$ and evaluate the integrals in (34.3) for obtaining the extremal distributions $P_{l,i,t,J}$ and $P_{u,i,t,J}$. Depending on the type of distributions $P^{i,t,J}$ we either have to compute a finite sum (in the case that $P^{i,t,J}$ is a sample, discrete with finite support respectively) or we have to employ efficient sampling techniques (see e.g. Deák 1990 [28]) for one- and two-dimensional integrals (in the case that $P^{i,t,J}$ is absolutely continuous or with support being too large in the number of realizations).

Remark 34.1: In the case that a sampled realization with positive probability is on the boundary of more than one sub-x-simplex we recommend to share that probability to those subsimplices; these can be controlled easily via the associated (zero and nonzero) generalized barycentric weights. In the case that some probabilities in (34.4) are 0, then we may choose any point within the subsimplex as atom (see Theorems 14.3 and 14.3'). Finally, again (34.4) illustrates the importance of a tricky usage of inverses and cross moment matrices to keep the computational operations low.

Having determined extremal distributions $P_{l,i,t,J}$ and $P_{u,i,t,J}$ (i=1,...,I) we next focus on extremal distributions $P_{l,t,J}$, $P_{u,t,J}$ corresponding to $P^{t,J}$ (the conditional distribution of P with respect to $\Omega^{t,J}$). $P_{l,t,J}$ and $P_{u,t,J}$ are achieved according to

$$P_{l,t,J} = \times_i \; P_{l,i,t,J} \quad (34.5.1)$$
$$P_{u,t,J} = \times_i \; P_{u,i,t,J} \,. \quad (34.5.2)$$

For reasons of completeness atoms and probabilities are obtained for $P_{l,t,J}$ with $\nu=(\nu_1,\ldots,\nu_I)$, $\nu_i=0,\ldots,K_i$, $i=1,\ldots,I$:

$$(a_\nu^{t,J}, \xi_\nu^{t,J}) = \times_i (a_{\nu_i}^{r(i,t,J)}, \xi_{\nu_i}^{s(i,t,J)}) \quad (34.6.1)$$

$$p_\Theta^{t,J}(\nu) = \prod_{i=1}^{I} p_\Theta^{r(i,t,J)}(\nu_i) \quad (34.6.2)$$

and for $P_{u,t,J}$ with $\mu=(\mu_1,\ldots,\mu_I)$, $\mu_i=0,\ldots,L_i$, $i=1,\ldots,I$:

$$(\eta_\mu^{t,J}, b_\mu^{t,J}) = \times_i (\eta_{\mu_i}^{r(i,t,J)}, b_{\mu_i}^{s(i,t,J)}) \quad (34.7.1)$$

$$p_\Xi^{t,J}(\mu) = \prod_{i=1}^{I} p_\Xi^{s(i,t,J)}(\mu_i) \;. \quad (34.7.2)$$

Finally, the desired extremal (lower and upper distributions) $P_{l,J}$ and $P_{u,J}$ for $\times_i(\Theta^i\times\Xi^i)$ are given with

$$P_{l,J} = \Sigma_t \,[P_{l,t,J} \,.\, \Pi_i \, P^i(\Theta^{r(i,t,J)}\times\Xi^{s(i,t,J)})] \quad (34.8.1)$$
$$P_{u,J} = \Sigma_t \,[P_{u,t,J} \,.\, \Pi_i \, P^i(\Theta^{r(i,t,J)}\times\Xi^{s(i,t,J)})] \;. \quad (34.8.2)$$

This yields in particular atoms and probabilities for $P_{l,J}$ with $t=1,\ldots,t^J$:

$$(a_\nu^{t,J}, \xi_\nu^{t,J}) = \times_i (a_{\nu_i}^{r(i,t,J)}, \xi_{\nu_i}^{s(i,t,J)}) \quad (34.9.1)$$
$$p_\Theta^{l,t,J}(\nu) = p_\Theta^{t,J}(\nu).\, \Pi_i \, P^i(\Theta^{r(i,t,J)}\times\Xi^{s(i,t,J)}) \quad (34.9.2)$$

and analogously for $P_{u,J}$ with $t=1,\ldots,t^J$

$$(\eta_\mu^{t,J}, b_\mu^{t,J}) = \times_i (\eta_{\mu_i}^{r(i,t,J)}, b_{\mu_i}^{s(i,t,J)}) \quad (34.10.1)$$
$$p_\Xi^{u,t,J}(\mu) = p_\Xi^{t,J}(\mu).\Pi_i \, P^i(\Theta^{r(i,t,J)}\times\Xi^{s(i,t,J)}) \,. \quad (34.10.2)$$

The cardinality of the supports of $P_{l,J}$ and $P_{u,J}$ are therefore given with

$$(\prod_{i=1}^{I} (K_i+1))*t^J \quad \text{and} \quad (\prod_{i=1}^{I} (L_i+1))*t^J \,. \quad (34.11)$$

We next state **ledd.dsc** and **uedd.dsc** for storing the extremal distributions $P_{l,J}$, $P_{u,J}$ in iteration J according to :

Name: LEDD.DSC

In the *Header* we specify

I	: number of subvectors being stochastically independent ;
K_i , L_i	: dimensions of the random vectors η^i, ξ^i , i=1,...I ;
nat	: total number of atoms (= $\Pi_i(K_i+1)*t^J$) - due to (34.11) .

In the *Body* we specify for each sub-x-simplex t=1,...,t^J and for each atom ν :

$p_\Theta^{l,t,J}(\nu)$	: probability ;
$(a_\nu^{t,J}, \xi_\nu^{t,J})$	: Σ_i (K_i +L_i) components of atom .

ENDDATA (ledd.dsc)

Name: UEDD.DSC

In the *Header* we specify

I	: number of subvectors being stochastically independent ;
K_i ,L_i	: dimensions of the random vectors η^i, ξ^i , i=1,...,I ;
nat	: total number of atoms (= $\Pi_i(L_i+1)*t^J$) - due to (34.11) .

In the *Body* we specify for each sub-x-simplex t=1,...,t^J and for each atom μ :

$p_\Xi^{u,t,J}(\mu)$	: probability ;
$(\eta_\mu^{t,J}, b_\mu^{t,J})$	: Σ_i (K_i +L_i) components of atom .

ENDDATA (uedd.dsc)

Due to (20.1) lower and upper extremal distributions ($P_{l,J}$ and $P_{u,J}$) of P subject to a given ×-simplicial partition $\mathcal{S}^J$ yield lower and upper approximating problems for (V.1) in iteration J :

$$(LAP)^J \qquad \inf_{x \in X} \; f(x) + \Sigma_t \Sigma_\nu \rho(a_\nu^{t,J}, \xi_\nu^{t,J}, x).p_\Theta^{l,t,J}(\nu) \qquad (35.1)$$

$$(UAP)^J \qquad f(x^{opt,J}) + \Sigma_t \Sigma_\mu \rho(\eta_\mu^{t,J}, b_\mu^{t,J}, x^{opt,J}).p_\Xi^{u,t,J}(\mu) \qquad (35.2)$$

where $x^{opt,J}$ denotes the optimal solution for $(LAP)^J$.

In the case that (V.1) is a mathematical program then again $(LAP)^J$ represents a blockstructured mathematical program and $(UAP)^J$ decomposes into $\Pi_i(K_i+1)*t^J$ independently solvable programs. The optimal solutions, subgradients respectively, and other information obtained for iteration J will be stored in *outap.sol* and *innap.sol* :

Name: OUTAP.SOL

In the *Header* we specify

nat	: number of atoms ($= \Pi_i (K_i+1)*t^J$) - according to (34.11) ;
n	: dimension of the first-stage decision x ;
$\Sigma_i K_i$	: dimension of $(\eta^1,...,\eta^I)$;
$\Sigma_i L_i$	: dimension of $(\xi^1,...,\xi^I)$;
$\Psi^{l,J}$	: optimal objective value of (35.1) ;
$x^{opt,J}$	: optimal solution (n components) of (35.1) .

In the *Body* we specify for each sub-×-simplex $t=1,...,t^J$ and for each atom ν :

$p_\Theta^{l,t,J}(\nu)$	: probability of atom ;
$\rho(a_\nu^{t,J},\xi_\nu^{t,J},x^{opt,J})$	: second-stage value ;
$\eta^*(a_\nu^{t,J},\xi_\nu^{t,J},x^{opt,J})$	: subgradient of $\rho(\eta,\xi,x^{opt,J})$ with respect to η at $(a_\nu^{t,J},\xi_\nu^{t,J})$;
$\xi^*(a_\nu^{t,J},\xi_\nu^{t,J},x^{opt,J})$	: subgradient of $\rho(\eta,\xi,x^{opt,J})$ with respect to ξ at $(a_\nu^{t,J},\xi_\nu^{t,J})$;
$(a_\nu^{t,J},\xi_\nu^{t,J})$	: subvector of atom ν .

ENDDATA(outap.sol)

Name: INNAP.SOL

In the *Header* we specify

nat	: number of atoms ($= \Pi_i (L_i+1)*t^J$) - according to (34.11) ;
n	: dimension of the first-stage decision x ;
$\Sigma_i K_i$	: dimension of $(\eta^1,...,\eta^I)$;
$\Sigma_i L_i$	: dimension of $(\xi^1,...,\xi^I)$;
$\Psi^{u,J}$	: objective value of (35.2) at $x^{opt,J}$;
$x^{opt,J}$	: first-stage decision to which the upper bound refers .

In the *Body* we specify for each sub-x-simplex $t=1,...,t^J$ and for each atom μ :

$p_\Xi^{u,t,J}(\mu)$	: probability of atom ;
$\rho(\eta_\mu^{t,J},b_\mu^{t,J},x^{opt,J})$	: second-stage value ;
$\eta^*(\eta_\mu^{t,J},b_\mu^{t,J},x^{opt,J})$	: subgradient of $\rho(\eta,\xi,x^{opt,J})$ subject to η at $(\eta_\mu^{t,J}, b_\mu^{t,J})$;
$\xi^*(\eta_\mu^{t,J},b_\mu^{t,J},x^{opt,J})$	: subgradient of $\rho(\eta,\xi,x^{opt,J})$ subject to ξ at $(\eta_\mu^{t,J}, b_\mu^{t,J})$;
$(\eta_\mu^{t,J},b_\mu^{t,J})$	: subvector of atom μ ;

ENDDATA(innap.sol)

36 x-Simplicial Refinement

Due to section 19 we know that it is of importance to concentrate on those sub-x-simplices (being leaves in *partition tree*) subject to which lower and upper bounds for the conditional expectation of the integrand differ most; within these sub-x-simplices we have to concentrate on those edges subject to which the integrand is 'mostly' nonlinear, to obtain significant improvements in the iterations. We stress that *ledd.dsc* and *outap.sol* as well as *uedd.dsc* and *innap.sol* coincide in the number of records in the corresponding *body* and further note that the *body* of each of these data structures might be seen as being split into t^J segments, each having $\Pi_i(K_i+1)$ records, each having $\Pi_i(L_i+1)$ records respectively. For verifying the refinement strategy we simply have to proceed according to the following steps :

For evaluating lower and upper bounds for the conditional expectation of the integrand subject to the leaves $\Omega^{t,J}$, we read all $\Pi_i(K_i+1)$, $\Pi_i(L_i+1)$ records of segment 't' of *outap.sol* and *innap.sol* , and perform :

$$\psi^{l,t,J} = \Sigma_\nu \, \rho(a_\nu^{t,J}, \xi_\nu^{t,J}, x^{opt,J}).p_\Theta^{t,J}(\nu) \tag{36.1.1}$$

$$\psi^{u,t,J} = \Sigma_\mu \, \rho(\eta_\mu^{t,J}, b_\mu^{t,J}, x^{opt,J}).p_\Xi^{t,J}(\mu) \quad . \tag{36.1.2}$$

In the case that for leaf t associated with for sub-×-simplex $\Omega^{t,J}$, the error exceeds our tolerance ε,

$$\psi^{u,t,J} - \psi^{l,t,J} > \varepsilon \quad , \tag{36.2}$$

we change the sign of the pointer of leaf t in *leaf.pot* and so characterize the set of sub-×-simplices, denoted E_ε^J, going to be partitioned in iteration J .
Having an arbitrary sub-×-simplex in E_ε^J, we check the degree of nonlinearity of the integrand on this sub-×-simplex, concentrate on $\Omega^{t,J} = \times_i (\Theta^{r(i,t,J)} \times \Xi^{s(i,t,J)})$ and consider all pairs of adjacent vertices of $\times_i \Theta^{r(i,t,J)}$ and of $\times_i \Xi^{s(i,t,J)}$.

We start with $\times_i \Theta^{r(i,t,J)}$ and define the set of adjacent vertices $\mathcal{E}^{l,t,J}$:

$$\mathcal{E}^{l,t,J} := \{ (\bar{v}, \hat{v}) \, / \, \exists \, i_1 : \bar{v}_{i_1} \neq \hat{v}_{i_1} \text{ and } \forall \, i \neq i_1 : \bar{v}_i = \hat{v}_i \;\; (\bar{v}_i, \hat{v}_i = 0,\ldots,K_i \,) \, \} \, . \tag{36.3}$$

We stress that there is no need to consider all pairs out of $\Pi_i(K_i+1)$ vertices and to test whether (36.3) is fulfilled. Instead all the pairs of adjacent vertices $(\bar{v}, \hat{v})$, totally

$$\sum_{i'=1}^{I} \binom{K_{i'}+1}{2} \, [\, \prod_{i \neq i'} (K_i+1)] \tag{36.4}$$

satisfying (36.3) are exactly determined through the following procedure:

$$
\begin{array}{l}
\mathcal{E}^{l,t,J} := \emptyset \\
\text{for all } \; i_1 = 1,\ldots,I \\
\text{for all } \; \bar{v}_{i_1} = 0,\ldots,K_{i_1} - 1 \\
\quad \text{for all } \; \hat{v}_{i_1} = \bar{v}_{i_1} + 1,\ldots,K_{i_1} \\
\quad\quad \text{for all } \; i \in \{1,\ldots,I\} \backslash \{ \, i_1 \, \} \\
\quad\quad\quad \text{for all } \; \bar{v}_i = 0,\ldots,K_i \\
\quad\quad\quad\quad \hat{v}_i := \bar{v}_i \\
\quad\quad\quad\quad \mathcal{E}^{l,t,J} := \mathcal{E}^{l,t,J} \cup (\bar{v}, \hat{v}) \; .
\end{array}
\tag{36.5}
$$

$\mathcal{E}^{1,t,J}$ is the desired set, as any 2 vertices in a simplex are adjacent and so define an edge. Having such a pair $(\bar{v}, \hat{v})$ we next simply read the associated two records in the t-th segment of the body of *outap.sol*, given through the record numbers

$$recnr\,(t, \bar{v}) = [\prod_{i=1}^{I} (K_i+1)] \,.\, (t-1) + \sum_{i'=1}^{I} [\prod_{i=i'+1}^{I} (K_i+1)] \,.\, \bar{v}_{i'} \qquad (36.6.1)$$

$$recnr\,(t, \hat{v}) = [\prod_{i=1}^{I} (K_i+1)] \,.\, (t-1) + \sum_{i'=1}^{I} [\prod_{i=i'+1}^{I} (K_i+1)] \,.\, \hat{v}_{i'} \;. \qquad (36.6.2)$$

(36.6) yields pointers for $\eta^*(a_{\bar{v}}^{t,J},\xi_{\bar{v}}^{t,J},x^{opt,J})$, $\xi^*(a_{\bar{v}}^{t,J},\xi_{\bar{v}}^{t,J},x^{opt,J})$ and $\eta^*(a_{\hat{v}}^{t,J},\xi_{\hat{v}}^{t,J},x^{opt,J})$, $\xi^*(a_{\hat{v}}^{t,J},\xi_{\hat{v}}^{t,J},x^{opt,J})$ and therefore the information required for evaluating $\Delta(\bar{v},\hat{v},x^{opt,J})$ according to (19.4.1), indicating the degree of nonlinearity of $\rho(.,.,x^{opt,J})$ along the associated line segment $[(a_{\bar{v}}^{t,J},\xi_{\bar{v}}^{t,J}),(a_{\hat{v}}^{t,J},\xi_{\hat{v}}^{t,J})]$.

We continue with considering $\times_i \Xi^{s(i,t,J)}$ and define $\mathcal{E}^{2,t,J}$, its set of adjacent vertices :

$$\mathcal{E}^{2,t,J} := \{(\bar{\mu},\hat{\mu}) \,/\, \exists\, i_1 : \bar{\mu}_{i_1} \neq \hat{\mu}_{i_1} \text{ and } \forall\, i \neq i_1 : \bar{\mu}_i = \hat{\mu}_i \;\; (\bar{\mu}_i, \hat{\mu}_i = 0,\ldots,L_i\,)\} \,. \qquad (36.7)$$

Analogously to the above, all the pairs of vertices $(\bar{\mu},\hat{\mu})$, now totally

$$\sum_{i'=1}^{I} \binom{L_{i'}+1}{2} [\prod_{i\neq i'} (L_i+1)] \;, \qquad (36.8)$$

satisfying (36.7) are exactly determined through :

$$\begin{array}{l}
\mathcal{E}^{2,t,J} := \emptyset \\
\text{for all } i_1 = 1,\ldots,I \\
\quad \text{for all } \bar{\mu}_{i_1} = 0,\ldots,L_{i_1} - 1 \\
\quad\quad \text{for all } \hat{\mu}_{i_1} = \bar{\mu}_{i_1}+1,\ldots,L_{i_1} \\
\quad\quad\quad \text{for all } i \in \{1,\ldots,I\}\backslash\{i_1\} \\
\quad\quad\quad\quad \text{for all } \bar{\mu}_i = 0,\ldots,L_i \\
\quad\quad\quad\quad\quad \hat{\mu}_i := \bar{\mu}_i \\
\quad\quad\quad\quad\quad \mathcal{E}^{2,t,J} := \mathcal{E}^{2,t,J} \cup (\bar{\mu},\hat{\mu}) \,.
\end{array} \qquad (36.9)$$

Having such a tuple $(\bar{\mu},\hat{\mu})$ we simply read the associated two records in the t-th segment of the body of *innap.sol*, given through the record numbers

$$recnr\,(t,\bar{\mu}\,) = [\prod_{i=1}^{I}(L_i+1)\,]\,.\,(t-1) + \sum_{i'=1}^{I}[\prod_{i=i'+1}^{I}(L_i+1)]\,.\,\bar{\mu}_{i'} \qquad (36.10.1)$$

$$recnr\,(t,\hat{\mu}\,) = [\prod_{i=1}^{I}(L_i+1)\,]\,.\,(t-1) + \sum_{i'=1}^{I}[\prod_{i=i'+1}^{I}(L_i+1)]\,.\,\hat{\mu}_{i'}\,. \qquad (36.10.2)$$

Again, (36.10) provides pointers for subgradients $\eta^*(\eta_{\bar{\mu}}^{t,J},b_{\bar{\mu}}^{t,J},x^{opt,J})$, $\xi^*(\eta_{\bar{\mu}}^{t,J},b_{\bar{\mu}}^{t,J},x^{opt,J})$ and $\eta^*(\eta_{\hat{\mu}}^{t,J},b_{\hat{\mu}}^{t,J},x^{opt,J})$, $\xi^*(\eta_{\hat{\mu}}^{t,J},b_{\hat{\mu}}^{t,J},x^{opt,J})$. This enables to get $\Delta(\bar{\mu}\,,\hat{\mu},x^{opt,J})$ of (19.4.2) indicating the degree of nonlinearity of $\rho(.,.,x^{opt,J})$ on line $[(\eta_{\bar{\mu}}^{t,J},b_{\bar{\mu}}^{t,J},),(\eta_{\hat{\mu}}^{t,J},b_{\hat{\mu}}^{t,J})]$.

With having $\Delta(\bar{v}\,,\hat{v},x^{opt,J})$ and $\Delta(\bar{\mu}\,,\hat{\mu},x^{opt,J})$ it remains to compute (19.5) :

$$\max\,\{\Delta(\bar{v}\,,\,\hat{v},x^J)\,,\,\Delta(\bar{\mu}\,,\hat{\mu},x^J)\mid\forall(\bar{v}\,,\hat{v})\in\mathcal{E}^{1,t,J},\;\forall\,(\bar{\mu}\,,\hat{\mu}\,)\in\mathcal{E}^{2,t,J}\,\}\,. \qquad (36.11)$$

For the ease of exposition let the optimal pair of adjacent vertices in (36.11) be denoted with a triple (nsw(t,J),nrvt1(t,J),nrvt2(t,J)) where (t,J) refers to $\Omega^{t,J}$ and

nsw(t,J):=1 if $\mathcal{E}^{1,t,J}$ covers the optimal value in (36.11) ;
nsw(t,J):=2 if $\mathcal{E}^{2,t,J}$ covers the optimal value in (36.11) - else ;
nrvt1(t,J),nrvt2(t,J) vectors of I integers stating the vertex numbers w.r.t. its enumeration of (36.5) and (36.9); note that these two vectors differ in exactly one component - due to adjacency.

Using the notation in section 33 we update the record nd=nd(t,J) in *node.pot* associated with leaf t in iteration J (i.e. associated to sub-×-simplex $\Omega^{t,J}$) :

ns1(nd(t,J)) is left unchanged as the characterization subject to subsimplices within Θ of any node remains the same.
ns2(nd(t,J)) is left unchanged as the characterization subject to subsimplices within Ξ of any node remains the same.

nswitch(nd(t,J)) := nsw(t,J) - this is obvious due to construction.

ptsx(nd(t,J)) is determined by that component $i\in\{1,...,I\}$ subject to which the vectors nrvt1(t,J),nrvt2(t,J) have been chosen adjacent; as a consequence the [ptsx(nd(t,J))]-th component of ns1(nd(t,J)) - if nswitch(nd(t,J))=1 - or the [ptsx(nd(t,J))]-th component of ns2(nd(t,J)) - if nswitch(nd(t,J))=2 - points to the subsimplex to be partitioned within the underlying sub-×-simplex $\Omega^{t,J}$.

ptv1(nd(t,J)) is determined by the [ptsx(nd(t,J))]-th component of nrvt1(t,J) .
ptv2(nd(t,J)) is determined by the [ptsx(nd(t,J))]-th component of nrvt2(t,J) .

wt1(nd(t,J)),wt2(nd(t,J)) will be set to 0.5 by default .

nv3(nd(t,J)) is determined by the quantity of vertices stored in *vertx'i'j'.pot* where 'i' is equal to the character represented by ptsx(nd(t,J)) and 'j' is equal to the character represented by nswitch(nd(t,J)).

npt0(nd(t,J)) is left unchanged as the 'father' of any node remains fixed.
npt1(nd(t,J))=nnode+1 and represents the number of 'left son' within *node.pot.*
npt2(nd(t,J))=nnode+2 and represents the number of 'right son' within *node.pot.*

The above information suffices to partition the desired subsimplex within *simpx'i'1.pot* if nswitch(nd(t,J))=1 (within *simpx'i'2.pot* if nswitch(nd(t,J))=2, respectively) and to extend the corresponding *simplex.pot* and *vertex.pot* correspondingly; next we specify the two nodes that correspond to the 'left' and 'right son' of sub-×-simplex $\Omega^{t,J}$ within *node.pot.*

Left Son of $\Omega^{t,J}$ will be stored at nd=nnode+1 within *node.pot* according to :

ns1(nnode+1) is equal to ns1(nd(t,J)) if nswitch(nd(t,J))=2; in the case that nswitch(nd(t,J))=1 then **ns1(nnode+1)** coincides with ns1(nd(t,J)) with exception of the [ptsx(nd(t,J))]-th component; this component represents the number of the first of the two new subsimplices generated within *simpx'i'1.pot* where 'i' is ptsx(nd(t,J)).
ns2(nnode+1) is equal to ns2(nd(t,J)) if nswitch(nd(t,J))=1; in the case that nswitch(nd(t,J))=2 then **ns2(nnode+1)** coincides with ns2(nd(t,J)) with exception of the [ptsx(nd(t,J))]-th component; this component represents the number of the first of the two new subsimplices generated within *simpx'i'2.pot* where 'i' is ptsx(nd(t,J)).

nswitch(nnode+1) := 0; (the 'left son' of a leaf becomes a leaf).
ptsx(nnode+1) := 0 .
ptv1(nnode+1) := 0 .
ptv2(nnode+1) := 0 .

wt1(nnode+1),wt2(nnode+1) will be set to 0.5 by default .

nv3(nnode+1) = 0 .

npt0(nnode+1) = nd(t,J) representing the node of the father .
npt1(nnode+1) = 0 .
npt2(nnode+1) = 0 .

Right Son of $\Omega^{t,J}$ will be stored at nd=nnode+2 within *node.pot* according to :

ns1(nnode+2) is equal to ns1(nd(t,J)) if nswitch(nd(t,J))=2; in the case that nswitch(nd(t,J))=1 then **ns1(nnode+2)** coincides with ns1(nd(t,J)), with exception of the [ptsx(nd(t,J))]-th component; this component represents the number of the second of the two new subsimplices generated within *simpx'i'1.pot* where 'i' is ptsx(nd(t,J)).
ns2(nnode+2) is equal to ns2(nd(t,J)) if nswitch(nd(t,J))=1; in the case that nswitch(nd(t,J))=2 then **ns2(nnode+2)** coincides with ns2(nd(t,J)), with exception of the [ptsx(nd(t,J))]-th component; this component represents the number of the second of the two new subsimplices generated within *simpx'i'2.pot* where 'i' is ptsx(nd(t,J)).

nswitch(nnode+2) := 0; (the 'right son' of a leaf becomes a leaf).
ptsx(nnode+2) := 0 .
ptv1(nnode+2) := 0 .
ptv2(nnode+2) := 0 .

wt1(nnode+2),wt2(nnode+2) will be set to 0.5 by default .

nv3(nnode+2) = 0 .

npt0(nnode+2) = nd(t,J) representing the node of the father ;
npt1(nnode+2) = 0 .
npt2(nnode+2) = 0 .

With setting nnode=nnode+2 in the header of *node.pot* we finish the update of *node.pot* w.r.t. to the partition of $\Omega^{t,J}$. In the case that all leaves within E_ε^J are split, we complete the refinement procedure and again determine the new extremal distributions associated with the new partition $\mathcal{S}^{J+1}$ of $\Theta\times\Xi$.

VI Solving Stochastic Linear Two-Stage Problems (Numerical Results & Computational Experiences)

In this chapter we discuss computational results achieved with applying the barycentric approximation scheme (sections 14,16,19,20) for solving *stochastic linear two-stage problems* with (C1)-(C5) of section 17 supposed to hold.

SLP-IOR is a model management system for stochastic linear programs (SLP) and currently under development by Kall and Mayer 1991 [75]. This system allows to model main SLP type models and gives OR practitioners the possibility to invoke available solvers for the individual model types. The present version has been developed on an IBM PS/2 Model 70 computer with 4 MB storage capacity. Within that configuration SLP-IOR employs the modeling system GAMS (see Brooke, Kendrick and Meeraus 1988 [19]) offering easy access to powerful general-purpose solvers in linear and nonlinear programming. The solver library of SLP-IOR includes besides these general-purpose packages also solvers especially designed to solve stochastic linear programs. These solvers are connected to SLP-IOR by developing interfaces to GAMS.

Based on the data structures specified in chapter V, the software package BRAIN has been included into the solver-library of the first version of SLP-IOR within an interface for iterative approximation. In the present version BRAIN provides the user with barycentric approximations of an underlying linear two-stage stochastic program: in particular, the data in *ledd.dsc* (sections 31, 34) define the lower approximating problems in (20.3.1). These represent dual blockstructured linear programs being solved with the package GAMS/MINOS. Similarly, *uedd.dsc* define the integrals in (20.1.4) and require to solve a sequence of linear recourse problems. This is done with the program package XMP. Primal and dual solutions of these approximate problems are transmitted to BRAIN according to the data structures *outap.sol* and *innap.sol* specified in sections 32 and 35.

Remark VI.1: The above cited program packages MINOS and XMP are well established within mathematical programming. MINOS is a program system designed to solve linear and nonlinear programming problems (see A. Murtagh and M.A. Saunders 1983 [100]). Models created with GAMS are solved with a special adapted version of MINOS called GAMS/MINOS. XMP is a linear programming library developed by Marsten 1981 [87]. In general any available solver might be invoke. For problems that relate to stochastic programs we mention in particular : NDSP and MSLiP (both, extended L-shaped methods to the multi-stage setting) by Birge 1985 [10] and Gassmann 1990 [52], QDECOM designed to solve linear large-scale blockstructured programs (Ruszczyński 1986 [132]), the Lagrangian Finite-Generation Method due to Rockafellar and Wets 1986 [125] (implemented by King 1988 [81]) designed to solve linear-quadratic blockstructured programs, or the Progressive Hedging Algorithm due to Rockafellar and Wets 1991 [127] (implemented and applied in Mulvey and Vladimirou 1989 [97], 1991 [98]), designed to solve blockstructured convex mathematical programs.

For the ease of understanding, let us recall stochastic two-stage models with the details being needed below :

$$\inf_{x\in X} \ c'x + \int_{\Theta\times\Xi} \rho(\eta,\xi,x)dP(\eta,\xi) \qquad \text{(VI.1.1)}$$

with

$$\begin{aligned} \rho(\eta,\xi,x) &:= \inf \ q(\eta)'y \\ \text{s.t.:} \quad & Wy=h(\xi)-T(\xi).x \\ & y\geq 0 \ . \end{aligned} \qquad \text{(VI.1.2)}$$

$X=\{x|\ Ax\leq a,\ x\geq 0\}$ represents a convex polyhedral subset of $\mathbf{R}^n$ given by the matrix $A\in\mathbf{R}^{m\times n}$ and vector $a\in\mathbf{R}^m$; $W\in\mathbf{R}^{m'\times n'}$ is the fixed complete recourse matrix and suppose dual feasibility to hold in Θ, i.e.: $\{u|W'u\leq q(\eta)\}\neq\emptyset$ for all $\eta\in\Theta$; this ensures the recourse function to be finite on $\Theta\times\Xi$, for any x. Further, q,h and T are assumed to be linear affine in η and ξ, i.e.:

$$q(\eta) = q_0 + \sum_{k=1}^{K} q_k.\eta_k \ , \ h(\xi) = h_0 + \sum_{l=1}^{L} h_l\,.\xi_l \ , \ \ T(\xi) = T_0 + \sum_{l=1}^{L} T_l\,.\xi_l \qquad \text{(VI.3)}$$

with $q_k\in\mathbf{R}^{n'}$, $h_l\in\mathbf{R}^{m'}$, $T_l\in\mathbf{R}^{m'\times n}$ (k=0,...,K; l=0,...,L). We allow to have stochastic independences compatible with $\Theta\times\Xi:=\times_i(\Theta^i\times\Xi^i)$, $(\eta,\xi)=\times_i(\eta^i,\xi^i)$ where $\Theta\subset\mathbf{R}^K$, $\Xi\subset\mathbf{R}^L$, $\Theta^i\subset\mathbf{R}^{K_i}$, $\Xi^i\subset\mathbf{R}^{L_i}$ and $P=\times_i P^i$ and supp $P^i\subset\Theta^i\times\Xi^i$ with Θ^i, Ξ^i representing (closed and regular) simplices for i=1,...,I .

Below we distinguish between test problems (within the class of linear two-stage programs) that are known from the literature (section 37) and such that are randomly generated (section 38).

For describing the computational results of the test problems and their solutions we use three tableaus : *input tableau, solution tableau* and *performance tableau* .

The **input tableau** states

i) the dimension of the matrices A(mxn),T(m'xn),W(m'xn') ;
ii) the number of stochastic independent subvectors , i.e. I ;
iii) the dimensions of the independent random vectors, i.e. K_i , L_i (i=1,...,I) ;
iv) the number of stochastic entries ;
v) the parameters of the uniform , normal distributions respectively ;
vi) the sample size N with respect to which the distributions are sampled.

The **solution tableau** states

i) lower and upper bounds for the optimal objective value ;
ii) the achieved relative accuracy ε ;
iii) the basic components of ε-appoximates for the optimal first stage solution.

The **performance tableau** tells us

i) how many iterations are needed for ε-optimality ;
ii) number of atoms for lower and upper extremal distributions in iteration J ;
iii) the achieved relative accuracy in iteration J ;
iv) the relative change of first-stage approximates within subsequent iterations.

Remark VI.2: The motivation to outline the relative changes in the first-stage approximates is the following: if the relative change is 0 within some iterations we may conclude that the refinements of the $\times$-simplices have no impact on the approximate solutions; so, the impact concentrates rather on the expectation of the recourse function with respect to the unchanged approximates. As a consequence, this allows to judge whether the difficulty of solving a problem lies either in the minimization or in the integration part (or in both).

All considered problems are formulated in two versions only differing in the distributions: i) the random variable are supposed to be uniformly distributed within the $\times$-simplices; ii) the random vectors are distributed normally, and truncated with respect to the $\times$-simplices. In the latter case, the parameters (mean and covariance matrix) are chosen in such a way, that the $\times$-simplices cover about 95-99% of the entire mass 1. The motivation to solve each problem twice (only differing in the distributions) is simply to illustrate the changes of optimal values and solutions with respect to two different classical distributions. So, whenever we speak of normal distributions below, we actually think of their being truncated with respect to the chosen ×-simplices. Finally, we have to mention, that in #5 of (20.1.5) within BRAIN we haven partitioned NOT ALL sub-×-simplices of E_ε^J but only those subject to which the approximates are least accurate (relative to the accuracy within a particular iteration).

37 Testproblems from Literature

According to the notation in literature we consider below the following test problems of type (VI.1): LOUVEAUX, SCARG7, SCRS8, CEP1 and APL1P. These test problems are solved w.r.t. distributions given in 'input-tableau'.

i) Testproblem **LOUVEAUX** has been derived from an energy problem by F. Louveaux (see Louveaux and Smeers 1988 [84]) .

INPUT - LOUVEAUX			
A(2x4)	T(7x4)		W(7x16)
Number of stochastic independent subvectors : I=4			
$K_1=1,\ L_1=0$	$K_2=0,\ L_2=1$	$K_3=0,\ L_3=1$	$K_4=0,\ L_4=1$
η_1	ξ_1	ξ_2	ξ_3
U(43,51)	U(1,9)	U(0,8)	U(0,8)
N(47,1)	N(5,1.5)	N(4,1.75)	N(4,1.25)
$N_1=1000$	$N_2=1000$	$N_3=1000$	$N_4=1000$
4 Stochastic Entries		Sample Size $N=10^{12}$	

SOLUTION - LOUVEAUX		
	Uniform	Normal (truncated)
lower and upper bounds for optimal value	[482.965 , 486.618]	[480.857 , 484.839]
accuracy	0.8 %	0.8%
approximate for optimal first stage solution	x1 = 2.891 x2 = 5.568 x3 = 4.583 x4 = 9.185	x1 = 1.854 x2 = 5.764 x3 = 5.237 x4 = 7.768

PERFORMANCE LOUVEAUX (s.t. Uniform Distribution)				
Iteration J	number of sub-x-simplices in iteration J	number of atoms $(P_{l,J} / P_{u,J})$	accuracy in % in iteration J	changes in x : $\| x^J - x^{J-1} \|$ in %
1	1	2/8	443	-
2	2	4/16	381	10.1
3	3	6/24	187	32.2
4	5	10/40	69	9.4
5	8	16/64	31	15.9
6	12	24/96	9.7	26.6
7	15	30/120	4.2	26.3
8	20	40/160	2.3	8.8
9	27	54/216	1.9	12.4
10	36	72/288	1.1	3.7
11	55	110/440	0.8	11.1

We realize that the approximates (of 0.8% accuracy) obtained with respect to uniform and truncated normal distribution, differ rather significantly; whereas their objective values are similar. Due to the performance tableau we have to stress that the numerical effort for achieving 0.8% accuracy is considerably larger for the truncated normal. In this case, we need 91 sub-x-simplices, yielding 182 atoms for the associated lower extremal distribution, which in turn yields a lower approximating problem of size 1276x2916. For the uniform distribution it suffices to have 55 sub-x-simplices, yielding 110 atoms and an associated lower approximating problem of size 772x1764.

PERFORMANCE LOUVEAUX
(s.t. truncated Normal Distribution)

Iteration J	number of sub-x-simplices in iteration J	number of atoms $(P_{l,J} / P_{u,J})$	accuracy in % in iteration J	changes in x : $\| x^J - x^{J-1} \|$ in %
1	1	2/8	468	-
2	2	4/16	394	8.0
3	3	6/24	275	17.5
4	5	10/40	90.8	16.8
5	8	16/64	25.3	10.4
6	13	26/104	11.9	31.3
7	20	40/160	8.7	14.1
8	28	56/224	3.8	11.8
9	41	82/328	2.3	15.8
10	63	126/504	1.5	14.6
11	91	182/728	0.8	2.4

ii) Testproblems **SCARG7, SCRS8** are due to John Birge and represent stochastic versions of problems appearing in Ho and Loute 1981 [60] .

INPUT - SCARG7		
A(15x20)	T(38x20)	W(38x40)
Number of stochastic independent subvectors : I=3		
$K_1=0, L_1=1$	$K_2=0, L_2=1$	$K_3=0, L_3=1$
ξ_1	ξ_1	ξ_2
U(1400,2400)	U(1000,2800)	U(500,1200)
N(1900,200)	N(1900,250)	N(850,100)
N_1 =1000	N_2 =1000	N_3 =1000
3 Stochastic Entries	Sample Size $N=10^9$	

PERFORMANCE SCARG7 (s.t. Uniform and Normal Distribution)				
Iteration J	number of sub-x-simplices in iteration J	number of atoms $(P_{l,J} / P_{u,J})$	accuracy in % in iteration J	changes in x : $\| x^J - x^{J-1} \|$ in %
1	1	1/8	0.4	-

With having a look on the performance tableau of SCARG7 we realize, that this problem is easy to solve, as with having the first moments of the random components we already obtain an approximate solution of 0.4% accuracy. Due to the fact that the chosen distributions coincide in the first moments and refer to the same ×-simplex, we achieve almost the same optimal objective and solutions.

SOLUTION - SCARG7		
	Uniform	Normal (truncated)
lower and upper bounds for the optimal value	[-834422.3, -8303844.6]	[-834343.5, -830771.6]
accuracy	0.4 %	0.4%
approximate for optimal first stage solution	x2 = 158 x3 = 158 x5 = 158 x6 = 158 x7 = 1324 x8 = 2504 x9 = 1022 x10 = 450 x13 = 1600 x15 = 158 x16 = 158 x17 = 375 x18 = 279 x19 = 1909	x2 = 158 x3 = 158 x5 = 158 x6 = 158 x7 = 1315 x8 = 2504 x9 = 1030 x10 = 450 x13 = 1600 x15 = 158 x16 = 158 x17 = 375 x18 = 279 x19 = 1921

Problem **SCRS8** is again rather challenging :

INPUT - SCRS8		
A(28x37)	T(28x37)	W(28x38)
Number of stochastic independent subvectors : I=3		
K_1=0, L_1=1	K_2=0, L_2=1	K_3=0, L_3=1
ξ_1	ξ_1	ξ_2
U(-2, 2)	U(-0.2, 0.2)	U(-0.4, 0.3)
N(0, 0.8)	N(0, 0.08)	N(-0.05, 0.16)
N_1 =1000	N_2 =1000	N_3 =1000
3 Stochastic Entries	Sample Size N=109	

PERFORMANCE SCRS8 (s.t. Uniform Distribution)				
Iteration J	number of sub-x-simplices in iteration J	number of atoms $(P_{l,J} / P_{u,J})$	accuracy in % in iteration J	changes in x : $\| x^J - x^{J-1} \|$ in %
1	1	1/8	400	-
2	2	2/16	252	96.5
3	3	3/24	133	0.0
4	4	4/32	59	0.0
5	6	6/48	26	66.9
6	9	9/72	18	54.2
7	14	14/112	9.3	29.8
8	22	22/176	4.3	4.3
9	33	33/264	2.5	25.8
10	49	49/392	1.1	0.9
11	75	75/600	0.5	10.8

PERFORMANCE SCRS8 (s.t. truncated Normal Distribution)				
Iteration J	number of sub-x-simplices in iteration J	number of atoms $(P_{l,J} / P_{u,J})$	accuracy in % in iteration J	changes in x : $\| x^J - x^{J-1} \|$ in %
1	1	1/8	406	-
2	2	2/16	217	0.0
3	4	4/32	40	0.0
4	7	7/56	23	0.0
5	11	11/88	14	5.4
6	17	17/136	9.5	0.0
7	24	24/192	5.5	17.3
8	31	31/248	3.1	0.0
9	37	37/296	1.4	0.0
10	45	45/360	0.5	0.1

SOLUTION - SCRS8		
	Uniform	Normal (truncated)
lower and upper bounds for the optimal value	[1014.767 , 1019.874]	[438.936 , 441.134]
accuracy	0.5 %	0.5%
approximate for optimal first stage solution	x5 = 5.21 x12 = 1.04 x15 = 0.065 x16 = 0.48 x19 = 1.0 x24 = 0.499 x29 = 0.2 x34 = 0.998	x5 = 0.00 x12 = 0.00 x15 = 0.065 x16 = 0.22 x19 = 1.0 x24 = 0.631 x29 = 0.2 x34 = 0.126

As it becomes evident from the solution-tableau, not only the computational effort for obtaining the 0.5% solution, but also optimal solution and values strongly depend on the distribution.

Due to the performance tableau we realize, in case of uniformly distributed random vectors we require 11 iterations for generating 75 sub-x-simplices that finally yield an approximate solution of 0.5% accuracy . If the random vector is truncated-normally distributed we succeed in determining an approximate solution of the same accuracy within 10 iterations and 45 sub-x-simplices. Further, it is interesting to note, that subject to uniformly distributed random data, the approximate solution in iterations J=1,...,11 strongly differ; whereas with truncated-normally distributed data there is only a remarkable change in iteration 6 of 17.3 %.

iii) Testproblem **CEP1** is taken from Higle, Sen and Yakowitz 1990 [59] :

INPUT - CEP1		
A(9x8)	T(7x8)	W(7x15)
Number of stochastic independent subvectors : I=3		
$K_1=0, L_1=1$	$K_2=0, L_2=1$	$K_3=0, L_3=1$
ξ_1	ξ_1	ξ_2
U(0,3000)	U(0,3000)	U(0,3000)
N(1500,600)	N(1500,600)	N(1500,600)
$N_1=1000$	$N_2=1000$	$N_3=1000$
3 Stochastic Entries	Sample Size $N=10^9$	

PERFORMANCE CEP1
(s.t. Uniform Distribution)

Iteration J	number of sub-x-simplices in iteration J	number of atoms $(P_{l,J} / P_{u,J})$	accuracy in % in iteration J	changes in x : $\| x^J - x^{J-1} \|$ in %
1	1	1/8	1630	-
2	2	2/16	582	65.6
3	3	3/24	165	0.0
4	4	4/32	147	0.0
5	7	7/56	62	0.0
6	11	11/88	41	0.0
7	19	19/152	29	0.0
8	29	29/232	17	0.0
9	47	47/376	11	0.0
10	73	73/584	7.7	0.0
11	111	111/888	5.4	0.0

We realize that the first iteration has an enormous gap of 1630%. Further, after the second iteration there is no change in the approximate for the optimal first-stage solution. This indicates that from the point of view of optimization, one may easily get an approximate solution that is a vertex of the convex polytope X, which is assumed to be given by deterministic data, offering an accuracy of at least 5.4 %. The difficulty lies obviously in computing the expectation of the recourse function with respect to the underlying first-stage approximate.

In the case that the random data are truncated-normally distributed we observe a similar behaviour below; with the same computational effort we obtain an accuracy of 9.4 %. As it becomes apparent from the solution tableau of CEP1, there is coincidence in the approximate first-stage solution, but a remarkable difference in the objective value.

PERFORMANCE CEP1
(s.t. truncated Normal Distribution)

Iteration J	number of sub-x-simplices in iteration J	number of atoms $(P_{l,J} / P_{u,J})$	accuracy in % in iteration J	changes in x : $\| x^J - x^{J-1} \|$ in %
1	1	1/8	1654	-
2	2	2/16	619	58.8
3	3	3/24	305	0.0
4	6	6/48	157	0.0
5	9	9/72	91	0.0
6	15	15/120	65	0.0
7	20	20/160	47	0.0
8	32	32/256	31	0.0
9	48	48/384	21	0.0
10	68	68/544	14	0.0
11	104	104/832	9.4	0.0

SOLUTION - CEP1		
	Uniform	Normal (truncated)
lower and upper bounds for the optimal value	[179321 , 189034]	[106079 , 116004]
accuracy	5.4 %	9.4%
approximate for optimal first stage solution	x3 = 2500 x4 = 2500 x6 = 250 x7 = 3000 x8 = 3000	x3 = 2500 x4 = 2500 x6 = 250 x7 = 3000 x8 = 3000

iv) Testproblem **APL1P** stems from Infanger 1991 [63]:

INPUT - APL1P		
A(2x2)	T(5x4)	W(5x9)
Number of stochastic independent subvectors : I=5		
$K_1 = K_2 = 0$, $L_1 = L_2 = 1$ ξ_1 , ξ_2 U(0, 1.0) N(0.5, 0.16) $N_1 = N_2 = 1000$		$K_3 = K_4 = K_5 = 0$, $L_3 = L_4 = L_5 = 1$ ξ_3 , ξ_4 , ξ_5 U(900,1200) N(1050,60) $N_3 = N_4 = N_5 = 1000$
5 Stochastic Entries		Sample Size $N=10^{15}$

SOLUTION - APL1P		
	Uniform	Normal (truncated)
lower and upper bounds for the optimal value	[2792.86 , 2817.8]	[2722.97, 2742.94]
accuracy	0.9 %	0.7%
approximate for optimal first stage solution	x1 = 1000 x2 = 2632	x1 = 1000 x2 = 2241

PERFORMANCE APL1P
(s.t. Uniform Distribution)

Iteration J	number of sub-x-simplices in iteration J	number of atoms ($P_{l,J}$ / $P_{u,J}$)	accuracy in % in iteration J	changes in x : $\| x^J - x^{J-1} \|$ in %
1	1	1/32	17	-
2	2	2/64	2.8	46.1
3	3	3/96	2.1	5.8
4	5	5/160	1.5	0.0
5	9	9/288	0.9	7.6

PERFORMANCE APL1P
(s.t. truncated Normal Distribution)

Iteration J	number of sub-x-simplices in iteration J	number of atoms ($P_{l,J}$ / $P_{u,J}$)	accuracy in % in iteration J	changes in x : $\| x^J - x^{J-1} \|$ in %
1	1	1/32	17	-
2	2	2/64	3.9	13.2
3	3	4/128	2.1	8.7
4	6	6/192	1.3	0.9
5	10	10/320	0.7	1.3

GENSLP is a software package (see Kall and Keller 1983 [74]) for generating randomly stochastic linear programs with recourse. The initial version allows for specifying complete or simple recourse with right hand side and/or technology matrix stochastic. An advanced version (due to Kall, Keller and Mayer 1991) allows for generating matrices with eligible sparsity and with second-stage objective to be stochastic. Using the advanced version we generated 25 test problems, differing in the dimensions, in sparsity of matrices and in stochastic independences. Out of these we chose 6: CR2A3, CR2A4, CR2B3, CR2B6, CR2C2, CR2D2.
Denoting with Δ^n the n-dimensional unit simplex, $U(\Delta^K \times \Delta^L)$ stands for the uniform distribution within the $\times$-simplex $\Delta^K \times \Delta^L$. $N(B^M, \Sigma^M)$ denotes a truncated multi-dimensional normal distribution with the following property: the expectation B^M is the (classical) barycenter of the M-dimensional unit simplex and Σ^M a covariance matrix, whose coefficients are chosen so that about 95-99% of the mass 1 lies in the M-dimensional unit simplex, implying that at least 95-99% lies in $\Delta^K \times \Delta^L$ with K+L=M. This allows for investigating the impact on the performance of the method, if supp P is considerably smaller than $\Delta^K \times \Delta^L$.

i) **CR2A3** : The matrices A, T and W are generated with a sparsity of 50% ; the dimension of the random vector is 7 (=M) .

INPUT - CR2A3		
A(3x6)	T(5x6)	W(5x10)
Number of stochastic independent subvectors: I=3		
$K_1=1, L_1=1$	$K_2=1, L_2=2$	$K_3=1, L_3=1$
(η_1, ξ_1)	(η_2, ξ_2, ξ_3)	(η_3, ξ_4)
$U(\Delta^1 \times \Delta^1)$	$U(\Delta^1 \times \Delta^2)$	$U(\Delta^1 \times \Delta^1)$
$N(B^2, \Sigma^2)$	$N(B^3, \Sigma^3)$	$N(B^2, \Sigma^2)$
$N_1 = 2000$	$N_2 = 2500$	$N_3 = 2000$
30 Stochastic Entries	Sample Size $N=10^{10}$	

SOLUTION - CR2A3		
	Uniform	Normal (truncated)
lower and upper bounds for the optimal value	[435.24, 442.11]	[245.71 , 250.21]
accuracy	1.6 %	1.8%
approximate for optimal first stage solution	x2 = 1.282 x3 = 0.976 x4 = 0.382	x2 = 1.282 x3 = 0.976 x4 = 0.382

PERFORMANCE CR2A3 (s.t. Uniform Distribution)				
Iteration J	number of sub-x-simplices in iteration J	number of atoms ($P_{l,J}$ / $P_{u,J}$)	accuracy in % in iteration J	changes in x : $\| x^J - x^{J-1} \|$ in %
1	1	8/12	34	-
2	2	16/24	15	0.0
3	3	24/36	7.0	0.0
4	5	40/60	5.8	0.0
5	8	64/96	3.5	0.0
6	15	120/180	2.6	0.0
7	24	192/288	2.1	0.0
8	40	320/480	1.6	0.0

PERFORMANCE CR2A3 (s.t. truncated Normal Distribution)				
Iteration J	number of sub-x-simplices in iteration J	number of atoms $(P_{l,J} / P_{u,J})$	accuracy in % in iteration J	changes in x : $\| x^J - x^{J-1} \|$ in %
1	1	8/12	55	-
2	2	16/24	29	0.0
3	3	24/36	11	0.0
4	4	32/48	8.7	0.0
5	6	48/72	6.1	0.0
6	9	72/108	4.7	0.0
7	14	112/168	3.5	0.0
8	19	152/228	2.8	0.0
9	30	240/360	2.2	0.0
10	42	336/504	1.8	0.0

Considering the required effort for obtaining an approximate solution with 1.6%, 1.8% accuracy, respectively, we realize that with respect to the chosen (truncated) normal distribution, the method performs considerably less partitions in the iterations. This is due to the fact that the support of the truncated normal distribution lies in, say one half, of the x-simplex. This necessitates to start with such partitions that investigate where the mass is posititioned within the x-simplex.

Further, the approximates remain unchanged during the iterations; at the beginning, the information of the initial cross moments amounts to choose a vertex of the deterministic constraints as initial approximate for the optimal first-stage decision. Additional cross moment information in subsequent iterations confirm that vertex to be optimal, and therefore contribute rather to improvements in approximating the expectation value.

We had to stop at 40, at 42 sub-x-simplices, respectively: 40 sub-x-simplices yield 320 atoms for the lower extremal distribution (with respect to the supposed dimensions of random vectors). This entails a linear block-structured lower approximating problem of size 1603x3210 with a density of 50% within the blocks. This about is the limit size for problems solvable with GAMS-MINOS on a PS/2 Model 70 with 4 MB storage capacity. The solution process of other problems, that have not been solved within 1% accuracy, had been stopped due to the same reason.

ii) **CR2A4** : The matrices A, T and W are generated with a sparsity of 50% ; the dimension of the random vector is 10 (=M) .

INPUT - CR2A4		
A(3x6)	T(5x6)	W(5x10)
Number of stochastic independent subvectors : I=3		
$K_1=2$, $L_1=1$ (η_1, η_2, ξ_1) $U(\Delta^2\times\Delta^1)$ $N(B^4,\Sigma^4)$ $N_1=2000$	$K_2=2$, $L_2=2$ $(\eta_3, \eta_4, \xi_2, \xi_3)$ $U(\Delta^2\times\Delta^2)$ $N(B^4,\Sigma^4)$ $N_2=2500$	$K_3=1$, $L_3=2$ (η_5, ξ_4, ξ_5) $U(\Delta^1\times\Delta^2)$ $N(B^3,\Sigma^3)$ $N_3=2000$
30 Stochastic Entries	Sample Size $N=10^{10}$	

SOLUTION - CR2A4		
	Uniform	Normal (truncated)
lower and upper bounds for the optimal value	[73.315, 83.753]	[37.410,46.004]
accuracy	14%	23%
approximate for optimal first stage solution	x1 = 0.167 x2 = 1.256 x3 = 0.976 x4 = 0.388	x2 = 0.839 x3 = 1.156 x5 = 0.976 x6 = 0.408

PERFORMANCE CR2A4 (s.t. Uniform Distribution)				
Iteration J	number of sub-x-simplices in iteration J	number of atoms $(P_{l,J} / P_{u,J})$	accuracy in % in iteration J	changes in x : $\lVert x^J - x^{J-1} \rVert$ in %
1	1	18/18	128	-
2	2	36/36	47	17
3	3	54/54	36	1.0
4	5	90/90	28	6.8
5	9	162/162	22	1.0
6	15	270/270	18	0.4
7	24	432/432	14	5.7

PERFORMANCE CR2A4 (s.t. truncated Normal Distribution)				
Iteration J	number of sub-x-simplices in iteration J	number of atoms $(P_{l,J} / P_{u,J})$	accuracy in % in iteration J	changes in x : $\lVert x^J - x^{J-1} \rVert$ in %
1	1	18/18	122	-
2	2	36/36	101	0.0
3	3	54/54	85	0.0
4	4	72/72	79	0.0
5	5	90/90	67	1.8
6	7	126/126	51	1.0
7	11	198/198	41	0.1
8	17	306/306	31	0.3
9	25	450/450	21	0.2

iii) **CR2B3** : The matrices A, T and W are generated with a sparsity of 20% ; the dimension of the random vector is 7 (=M) .

INPUT - CR2B3		
A(5x10)	T(10x10)	W(10x20)
Number of stochastic independent subvectors : I=3		
$K_1=1$, $L_1=1$	$K_2=1$, $L_2=2$	$K_3=1$, $L_3=1$
(η_1, ξ_1)	(η_2, ξ_2, ξ_3)	(η_3, ξ_4)
$U(\Delta^1 \times \Delta^1)$	$U(\Delta^1 \times \Delta^2)$	$U(\Delta^1 \times \Delta^1)$
$N(B^2, \Sigma^2)$	$N(B^3, \Sigma^3)$	$N(B^2, \Sigma^2)$
$N_1 = 2000$	$N_2 = 2500$	$N_3 = 2000$
50 Stochastic Entries		Sample Size $N=10^{10}$

SOLUTION - CR2B3		
	Uniform	Normal (truncated)
lower and upper bounds for the optimal value	[285.33 , 299.74]	[159.02 , 170.29]
accuracy	5.0 %	7.1%
approximate for optimal first stage solution	x1 = 0.513 x2 = 0.075 x3 = 0.830 x4 = 0.873 x5 = 0.935 x6 = 1.094 x9 = 0.537 x10= 0.011	x1 = 0.397 x2 = 0.075 x3 = 0.830 x4 = 0.718 x5 = 0.935 x6 = 1.043 x9 = 0.670 x10 = 0.331

PERFORMANCE CR2B3 (s.t. Uniform Distribution)				
Iteration J	number of sub-x-simplices in iteration J	number of atoms $(P_{l,J} / P_{u,J})$	accuracy in % in iteration J	changes in x : $\|\| x^J - x^{J-1} \|\|$ in %
1	1	8/12	37	-
2	2	16/24	16	17
3	3	24/36	13	8.3
4	5	40/60	11	6.9
5	8	64/96	8.6	1.9
6	12	96/144	6.7	2.2
7	21	168/252	5.0	1.5

PERFORMANCE CR2B3 (s.t. truncated Normal Distribution)				
Iteration J	number of sub-x-simplices in iteration J	number of atoms $(P_{l,J} / P_{u,J})$	accuracy in % in iteration J	changes in x : $\|\| x^J - x^{J-1} \|\|$ in %
1	1	8/12	50	-
2	2	16/24	44	0.0
3	3	24/36	32	0.0
4	4	32/48	26	0.5
5	5	40/60	21	0.0
6	6	48/72	17	0.0
7	7	56/84	16	0.3
8	8	64/96	15	0.4
9	10	80/120	13	0.9
10	14	112/168	9.3	1.3
11	19	152/228	7.1	0.4

iv) **CR2B6** : Matrices A,T and W are generated with a sparsity of 20 %; the dimension of the random vector is 10 (=M) .

INPUT - CR2B6		
A(5x10)	T(10x10)	W(10x20)
Number of stochastic independent subvectors : I=1 K_1=4, L_1=6 $(\eta_1, \eta_2, \eta_3, \eta_4, \xi_1, \xi_2, \xi_3, \xi_4, \xi_5, \xi_6)$ $U(\Delta^4 \times \Delta^6)$ $N(B^{10}, \Sigma^{10})$ N_1 =3000		
50 Stochastic Entries		Sample Size $N=3x10^3$

We want to stress, that for the supposed 10-dimensional probability space we manage to improve the accuracy within 37 partitions from 138% to 18 % for uniformly distributed random vectors; for the truncated-normally distributed random vectors, we require 38 partitions to improve the accuracy from 94% to 10 %. We had to stop as the lower approximating problem reached the limit with size 1855x3710, with size 1905x3810 respectively .

The same observation is valid for test problem CR2A4 : Within 24 we improved the accuracy from 128 % to 14 % for the uniform distribution; within 25 partitions the accuracy has been improved to 21 % having started with 122 % for truncated-normal distribution. Herewith, we observe some delay in case the random vectors are truncated-normally distributed: this is due to the fact, that the support of the truncated normal distribution lies within one half of the chosen ×-simplex; this necessitates first to investigate in which part of the ×-simplex the support of the distribution is situated.

SOLUTION - CR2B6		
	Uniformly	Normal (truncated)
lower and upper bounds for the optimal value	[200.25 , 236.02]	[118.64 , 131.03]
accuracy	18 %	10%
approximate for optimal first stage solution	x1 = 0.504 x2 = 0.435 x3 = 0.750 x4 = 0.123 x5 = 0.054 x6 = 0.082 x7 = 0.632 x9 = 0.558 x10 = 0.708	x1 = 0.277 x2 = 0.469 x3 = 0.750 x4 = 0.123 x5 = 0.255 x6 = 0.000 x7 = 0.455 x9 = 0.714 x10 = 0.790

PERFORMANCE CR2B6 (s.t. Uniform Distribution)				
Iteration J	number of sub-x-simplices in iteration J	number of atoms ($P_{l,J}$ / $P_{u,J}$)	accuracy in % in iteration J	changes in x : $\| x^J - x^{J-1} \|$ in %
1	1	5/7	138	-
2	2	10/14	90	12
3	4	20/28	46	14
4	8	40/56	29	10
5	14	70/98	27	4.5
6	23	115/161	22	3.0
7	37	185/259	18	1.0

PERFORMANCE CR2B6 (s.t. Normal Distribution)				
Iteration J	number of sub-x-simplices in iteration J	number of atoms $(P_{l,J} / P_{u,J})$	accuracy in % in iteration J	changes in x : $\| x^J - x^{J-1} \|$ in %
1	1	5/7	94	-
2	2	10/14	84	0.0
3	3	15/21	67	0.0
4	4	20/28	46	2.7
5	6	30/42	38	0.4
6	8	40/56	30	1.2
7	10	50/70	22	10
8	14	70/98	16	0.2
9	20	100/140	14	0.7
10	28	140/196	11	0.8
11	38	190/266	10	0.4

v) **CR2C2**: The matrices A, T and W are generated with a sparsity of 15% ; the dimension of the random vector is 5 (=M) .

INPUT - CR2C2		
A(10x20)	T(15x20)	W(15x25)
Number of stochastic independent subvectors : I=3		
$K_1=1, L_1=2$ (η_1, ξ_1, ξ_2) $U(\Delta^1 \times \Delta^2)$ $N(B^3, \Sigma^3)$ $N_1=3000$		$K_2=1, L_2=1$ (η_2, ξ_3) $U(\Delta^1 \times \Delta^1)$ $N(B^2, \Sigma^2)$ $N_2=2000$
85 Stochastic Entries		Sample Size $N=6 \times 10^6$

SOLUTION - CR2C2		
	Uniform	Normal (truncated)
lower and upper bounds for the optimal value	[340.07 , 345.41]	[253.78 , 261.17]
accuracy	1.6%	2.9%
approximate for optimal first stage solution	x1 = 0.781 x2 = 0.173 x3 = 0.568 x4 = 0.000 x5 = 0.017 x6 = 1.044 x7 = 0.589 x9 = 0.723 x11 = 0.000 x12 = 0.990 x13 = 0.084 x14 = 0.301 x15 = 0.556 x16 = 1.171 x17 = 0.262 x19 = 0.092	x1 = 0.722 x2 = 0.330 x3 = 0.577 x4 = 0.282 x5 = 0.017 x6 = 1.044 x7 = 0.589 x9 = 0.723 x11 = 0.437 x12 = 0.875 x13 = 0.000 x14 = 0.605 x15 = 0.738 x16 = 1.015 x17 = 0.056 x19 = 0.000

PERFORMANCE CR2C2 (s.t. Uniform Distribution)				
Iteration J	number of sub-x-simplices in iteration J	number of atoms ($P_{l,J}$ / $P_{u,J}$)	accuracy in % in iteration J	changes in x : $\| x^J - x^{J-1} \|$ in %
1	1	4/6	92	-
2	2	8/12	25	26
3	4	16/24	9.6	11
4	6	24/36	5.9	9.5
5	11	44/66	4.2	3.3
6	17	68/102	2.2	4.1
7	26	104/156	1.6	0.8

PERFORMANCE CR2C2 (s.t. truncated Normal Distribution)				
Iteration J	number of sub-x-simplices in iteration J	number of atoms $(P_{l,J} / P_{u,J})$	accuracy in % in iteration J	changes in x : $\| x^J - x^{J-1} \|$ in %
1	1	4/6	80	-
2	2	8/12	36	25
3	4	16/24	25	0.4
4	6	24/36	14	11
5	9	36/54	8.4	15
6	13	52/78	7.0	2.1
7	17	68/102	6.3	0.2
8	22	88/132	3.5	0.9
9	30	120/180	2.9	0.5

vi) **CR2D2** : The matrices A, T and W are generated with a sparsity of 10%; the dimension of the random vector is 5 (=M) .

INPUT - CR2D2		
A(15x25)	T(20x25)	W(20x30)
Number of stochastic independent subvectors : I=2		
$K_1=1, L_1=2$ (η_1, ξ_1, ξ_2) $U(\Delta^1 \times \Delta^2)$ $N(B^3, \Sigma^3)$ $N_1=3000$		$K_2=1, L_2=1$ (η_2, ξ_3) $U(\Delta^1 \times \Delta^1)$ $N(B^2, \Sigma^2)$ $N_2=2000$
100 Stochastic Entries		Sample Size $N=6\text{x}10^6$

PERFORMANCE CR2D2 (s.t. Uniform Distribution)				
Iteration J	number of sub-x-simplices in iteration J	number of atoms $(P_{l,J} / P_{u,J})$	accuracy in % in iteration J	changes in x : $\| x^J - x^{J-1} \|$ in %
1	1	4/6	61	-
2	2	8/12	25	7.4
3	4	16/24	13	40
4	7	28/42	8.7	18
5	12	48/72	4.9	6.6
6	19	76/114	3.1	5.8

PERFORMANCE CR2D2 (s.t. truncated Normal Distribution)				
Iteration J	number of sub-x-simplices in iteration J	number of atoms $(P_{l,J} / P_{u,J})$	accuracy in % in iteration J	changes in x : $\| x^J - x^{J-1} \|$ in %
1	1	4/6	54	-
2	2	8/12	31	0.0
3	3	12/18	17	1.1
4	5	20/30	9.1	0.0
5	7	28/42	3.7	1.0
6	9	36/54	3.1	0.0
7	11	44/66	2.1	0.3
8	14	56/84	1.4	0.0
9	18	72/108	1.0	0.0
10	24	96/144	0.5	0.6

SOLUTION - CR2D2		
	Uniform	Normal (truncated)
lower and upper bounds for the optimal value	[1331.11 , 1372.30]	[974.89 , 980.13]
accuracy	3.1%	0.5%
approximate for optimal first stage solution	x1 = 0.842 x2 = 0.377 x3 = 0.180 x4 = 0.561 x5 = 0.180 x6 = 0.281 x7 = 0.000 x8 = 0.000 x9 = 0.203 x10 = 1.685 x11 = 0.158 x12 = 0.581 x13 = 0.000 x14 = 0.260 x15 = 0.000 x16 = 0.000 x17 = 0.568 x18 = 2.492 x19 = 0.000 x20 = 2.911 x21 = 0.547 x22 = 1.236 x23 = 0.000 x24 = 0.208 x25 = 0.889	x1 = 0.842 x2 = 0.000 x3 = 0.180 x4 = 0.518 x5 = 0.180 x6 = 0.281 x7 = 0.000 x8 = 0.000 x9 = 0.000 x10 = 1.685 x11 = 0.000 x12 = 0.581 x13 = 0.000 x14 = 0.260 x15 = 0.000 x16 = 0.139 x17 = 0.568 x18 = 4.448 x19 = 0.000 x20 = 3.783 x21 = 0.547 x22 = 4.177 x23 = 0.000 x24 = 0.208 x25 = 0.073

Finally, we report on computing times for evaluating the extremal distributions, for solving the block structured lower approximating problem (20.1.3) and for solving the sequence of recourse problems (20.1.4). The block structured program is solved with GAMS/MINOS ; we have to stress that this solver does not exploit the features of unchanged data within the blocks; for us there has not been available an alternative (competitive) package within the configuration IBM PS/2

(4 MB, 20MHz) and with accessing the extended memory. So below, the computing times for solving the block structured problems may be seen as upper bounds.

Obviously, computing times strongly depend on the dimensions of the probability space, on the dimensions of the matrices and on the number of iterations needed to get the approximate solution. Taking this dependency into account, with respect to the entire battery of testproblems, there have been no significant differences in computing times. We therefore restrict on one randomly generated testproblem, chosen **CR2B3**, for which we outline and discuss the computing times, needed for *evaluating extremal distributions* (with BRAIN), *solving lower approximate problems* (with GAMS/MINOS) and *solving the sequences of recourse problems* (with XMP).

Computing Times (CR2B3 - truncated Normal Distribution)

number of sub-x-simplices	number of atoms of P_l	number of atoms of P_u	BRAIN	GAMS/ MINOS	XMP
1	8	12	0'38"	0'21"	0'18
2	16	24	1'38"	0'42"	0'23
3	24	36	2'19"	1'28"	0'26
4	32	48	3'03"	2'15"	0'30
5	40	60	3'45"	3'08"	0'34
6	48	72	4'26"	4'33"	0'38
7	56	84	5'07"	4'55"	0'43"
8	64	96	5'49"	7'29"	0'46"
10	80	120	7'11"	13'54"	0'55"
14	112	168	9'52"	22'25"	1'13"
19	152	228	13'11"	43'29"	1'34"

Above computing times illustrate that BRAIN behaves linearly with respect to the number of sub-x-simplices which define the partition. Even a better behavior is observable with applying XMP to a sequence of recourse problems, as can be expected due to the fact that the recourse problems are to solve independently and due to the possibility to start with a good initial solution. As wellknown too, computing times increase rather strongly with the size of the linear problem for GAMS/MINOS (and of course for any other general-purpose LP-solver, too).

Remark 38.1 : The computer times observed with BRAIN do not include the time, necessary for performing the refinement scheme (20.1.7). As refinement is done within seconds in all iterations, we have realized that these computing times are in fact negligible.

Let us discuss values and sizes of the first and last iteration :
In iteration 1 we have one sub-×-simplex, P_l consisting of 8 atoms and P_u consisting of 12 atoms. For evaluating these extremal distributions BRAIN requires 38 seconds. The lower approximating problem consists of 8 blocks (with a sparsity of 25%) and is of size 85x170. GAMS/MINOS determines the optimal solution of that problem in 21 seconds. XMP has to solve 12 linear problems with different right-hand sides and objectives. This is done in 18 seconds .
Up to iteration 11 we had split the initial ×-simplex into 19 sub-×-simplices; this entails P_l to consist of 152 atoms and P_u to consist of 228 atoms. These extremal distributions are computed by BRAIN within about 13 minutes. The associated lower approximating problem has 152 blocks (with the same sparsity of 25% each) and has so a total size of 1525x3045. For solving this LP, MINOS/GAMS requires (from scratch) almost 44 minutes. XMP has to solve 228 recourse problems; this is done in 1'34" with using hot starts.

For reasons of completeness we state also the corresponding computing times of those problems from literature which have been solved in section 37.

Computing Times (SCARG7 - Uniform & truncated Normal Distribution)

number of sub-x-simplices	number of atoms of P_l	number of atoms of P_u	BRAIN	GAMS/ MINOS	XMP
1	1	8	0'25"	0'32"	0'14"

Computing Times (LOUVEAUX - Uniform Distribution)

number of sub-x-simplices	number of atoms of P_l	number of atoms of P_u	BRAIN	GAMS/ MINOS	XMP
1	2	8	0'31"	0'02"	0'16"
2	4	16	0'49"	0'04"	0'18"
3	6	24	1'06"	0'05"	0'20"
5	10	40	1'40"	0'09"	0'26"
8	16	64	2'29"	0'14"	0'33"
12	24	96	3'35"	0'23"	0'44"
15	30	120	4'23"	0'34"	0'51"
20	40	160	5'44"	0'52"	1'05"
27	54	216	7'38"	1'22"	1'23"
36	72	288	10'02"	2'19"	1'47"
55	110	440	15'09"	5'39"	2'37"

Computing Times (LOUVEAUX - truncated Normal Distribution)

number of sub-x-simplices	number of atoms of P_l	number of atoms of P_u	BRAIN	GAMS/ MINOS	XMP
1	2	8	0'32"	0'03"	0'16"
2	4	16	0'50"	0'04"	0'18"
3	6	24	1'06"	0'05"	0'20"
5	10	40	1'40"	0'08"	0'25"
8	16	64	2'29"	0'14"	0'33"
13	26	104	3'51"	0'27"	0'46"
20	40	160	5'46"	0'51"	1'03"
28	56	224	7'54"	1'50"	1'25"
41	82	328	11'23"	3'42"	1'57"
63	126	504	17'17"	8'34"	2'53"
91	182	728	24'43"	19'19"	4'06"

Computing Times (SCRS8 - Uniform Distribution)

number of sub-x-simplices	number of atoms of P_l	number of atoms of P_u	BRAIN	GAMS/ MINOS	XMP
1	1	8	0'28"	0'03"	0'24"
2	2	16	0'40"	0'07"	0'34"
3	3	24	0'52"	0'10"	0'43"
4	4	32	1'04"	0'15"	0'52"
6	6	48	1'28"	0'24"	1'11"
9	9	72	2'04"	0'43"	1'40"
14	14	112	3'05"	1'13"	2'26"
22	22	176	4'39"	2'58"	3'38"
33	33	264	6'50"	5'35"	5'26"
49	49	392	9'54"	12'09"	7'50"
75	75	600	14'59"	28'54"	12'07"

Computing Times (SCRS8 - truncated Normal Distribution)

number of sub-x-simplices	number of atoms of P_l	number of atoms of P_u	BRAIN	GAMS/ MINOS	XMP
1	1	8	0'27"	0'04"	0'25"
2	2	16	0'40"	0'08"	0'33"
4	4	32	1'04"	0'16"	0'51"
7	7	56	1'39"	0'29"	1'17"
11	11	88	2'27"	0'52"	1'54"
17	17	136	3'38"	1'26"	2'48"
24	24	192	5'02"	2'40"	3'53"
31	31	248	6'24"	4'05"	5'01"
37	37	296	7'34"	5'17"	5'59"
45	45	360	8'42"	8'02"	7'18"

Computing Times (CEP1 - Uniform Distribution)

number of sub-x-simplices	number of atoms of P_l	number of atoms of P_u	BRAIN	GAMS/ MINOS	XMP
1	1	8	0'28"	0'02"	0'15"
2	2	16	0'39"	0'04"	0'17"
3	3	24	0'52"	0'05"	0'19"
4	4	32	1'05"	0'06"	0'21"
7	7	56	1'40"	0'08"	0'28"
11	11	88	2'27"	0'11"	0'37"
19	19	152	4'00"	0'18"	0'54"
29	29	232	5'54"	0'33"	1'14"
47	47	376	9'20"	1'14"	1'55"
73	73	584	14'15"	2'18"	2'53"
111	111	888	21'21"	4'59"	4'14"

Computing Times (CEP1 - truncated Normal Distribution)

number of sub-x-simplices	number of atoms of P_l	number of atoms of P_u	BRAIN	GAMS/ MINOS	XMP
1	1	8	0'27"	0'03"	0'16"
2	2	16	0'39"	0'04"	0'18"
3	3	24	0'52"	0'05"	0'20"
6	6	48	1'28"	0'06"	0'27"
9	9	72	2'03"	0'09"	0'32"
15	15	120	3'12"	0'15"	0'44"
20	20	160	4'12"	0'22"	0'54"
32	32	256	6'29"	0'42"	1'22"
48	48	384	9'33"	1'22"	2'04"
68	68	544	13'17"	2'20"	2'44"
104	104	832	20'04"	4'40"	4'06"

Computing Times (APL1P - Uniform Distribution)					
number of sub-x-simplices	number of atoms of P_l	number of atoms of P_u	BRAIN	GAMS/ MINOS	XMP
1	1	32	0'37"	0'03"	0'19"
2	2	64	1'00"	0'04"	0'25"
3	3	96	1'22"	0'05"	0'31"
5	5	160	2'08"	0'07"	0'43"
9	9	288	3'37"	0'08"	1'06"

Computing Times (APL1P - truncated Normal Distribution)					
number of sub-x-simplices	number of atoms of P_l	number of atoms of P_u	BRAIN	GAMS/ MINOS	XMP
1	1	32	0'37"	0'03"	0'19"
2	2	64	1'01"	0'05"	0'25"
3	3	128	1'46"	0'06"	0'36"
6	6	192	2'32"	0'07"	0'48"
10	10	320	4'01"	0'08"	1'11"

The above results strongly document the importance of solving the sequence of lower approximating problems efficiently. It is an obvious fact that above computing times can strongly be reduced if those features are exploited that characterize mathematical programs arising from approximating stochastic programs: in particular, blockstructure, unchanged data within the blocks. Moreover, we have to stress that the better the approximation, the better the associated approximate solution serves as initial starting point for the next iteration, where the lower approximating problem might be considerably larger than in the previous iteration. Nevertheless, this fact could likely entail that the subsequent larger problem might be solved even faster than the (previous) smaller problem.

BIBLIOGRAPHY

[1] H. ATTOUCH and R.J.-B. WETS : Approximation and Convergence in Nonlinear Optimization; in : O. Mangasarian, R. Meyer and S.M. Robinson (eds.) : Nonlinear Programming 4 ; Academic Press (1981) 367-394

[2] M. AVRIEL and A.C. WILLIAMS : The Value of Information and Stochastic Programming ; Operations Research 18 (1970) 947-954

[3] H. BAUER : Wahrscheinlichkeitstheorie (4. Auflage) ; W. de Gruyter 1990

[4] M.S. BAZARAA and C.M. SHETTY : Nonlinear Programming, Theory and Algorithms; John Wiley & Sons 1979

[5] E. M. BEALE : On Minimizing a Convex Function subject to Linear Inequalities; J. Royal Stat. Soc. 17b (1955) 173-184

[6] J.F. BENDERS : Partitioning Procedures for Solving Mixed-Variable Programming Problems; Numerische Mathematik 4 (1962) 238-252

[7] A. BEN-TAL and E. HOCHMAN : More Bounds on the Expectation of a Random Variable; Journal of Appl. Prob. 9 (1972) 803-812

[8] D. BERTSEKAS : Constrained Optimization and Lagrange Multiplier Methods; Academic Press 1982

[9] P. BILLINGSLEY : Convergence of Probability Measures; John Wiley & Sons 1968

[10] J. BIRGE : Decomposition and Partitioning Methods for Multi-Stage Stochastic Linear Programs; Operations Research 33 (1985) 989-1007

[11] J. BIRGE : An L-shaped Computer Code for Multi-Stage Stochastic Linear Programs; in: Y. Ermoliev and R.J.-B. Wets (eds.): Numerical Techniques for Stochastic Optimization; Springer-Verlag (1988) 255-266

[12] J. BIRGE : The Relationship Between the L-shaped Method and Dual Basis Factorization for Stochastic Linear Programming; in: Y. Ermoliev and R.J.-B. Wets (eds.): Numerical Techniques for Stochastic Optimization; Springer-Verlag (1988) 267- 272

[13] J. BIRGE and M. TEBOULLE : Upper Bounds on the Expected Value of a Convex Function using Gradient and Conjugate Information; Math. of OR 14/4 (1989) 745-759

[14] J. BIRGE and R.J.-B. WETS : Designing Approximation Schemes for Stochastic Optimization Problems, in Particular for Stochastic Programs with Recourse; Math. Progr. Study 27 (1986) 54-102

[15] J. BIRGE and R.J.-B. WETS : Computing Bounds for Stochastic Programming Problems by Means of a Generalized Moment Problem; Math. of OR 12 (1987) 149-162

[16] J. BIRGE and R.J.-B. WETS : Sublinear Upper Bounds for Stochastic Programming with Recourse; Math. Progr. 43 (1989) 131-149

[17] J. BIRGE and R.J.-B. WETS : Stochastic Programming, Part I,II; Proceedings of the 5th International Conference on Stochastic Programming, Ann Arbor, Michigan, August 13-18, 1989; Annals of Operations Research 30-31, 1991

[18] A. BRØNDSTED : An Introduction to Convex Polytopes; Springer-Verlag 1983

[19] A. BROOKE, D. KENDRICK and A. MEERAUS : GAMS. A User's Guide; The Scientific Press 1988

[20] C. CASTAING and M. VALADIER : Convex Analysis and Measurable Multifunctions ; Lecture Notes in Mathematics 580, Springer-Verlag 1977

[21] A. CHARNES and W.W. COOPER : Chance-Constrained Programming; Management Science 5 (1959) 197-207

[22] F. CLARKE : Optimization and Nonsmooth Analysis; Centre de Recherches Mathématiques, Université de Montréal 1989

[23] G.B. DANTZIG : Linear Programming under Uncertainty; Management Science 1 (1955) 197-206

[24] G.B. DANTZIG : Linear Programming and Extensions; Princeton University Press 1963

[25] G.B. DANTZIG and P.W. GLYNN : Parallel Processors for Planning under Uncertainty; Annals of Operations Research 22 (1990) 1-21

[26] G.B. DANTZIG, P.W. GLYNN, M. AVRIEL, J.C. STONE, R. ENTRIKEN and M. NAKAYAMA : Decomposition Techniques for Multiarea Generation and Transmission Planning under Uncertainty; Stanford University, Stanford CA 1989

[27] P.J. DAVIS and P. RABINOWITZ : Methods of Numerical Integration; Academic Press 1975

[28] I. DÉAK : Random Number Generators and Simulation; in: A. Prékopa (eds.) : Mathematical Methods of Operations Research ; Akadémiai Kiadó , Budapest 1990

[29] M.A.H. DEMPSTER (ed.): Stochastic Programming; Proceedings of the 1st International Conference on Stochastic Programming, Oxford 1974, Academic Press 1980

[30] M.A.H. DEMPSTER : Introduction to Stochastic Programming ; in : M.A.H. Dempster (ed.): Stochastic Programming; Academic Press (1980) 3-59

[31] M.A.H. DEMPSTER : On Stochastic Programming II: Dynamic Problems under Risk; Research Report DAL TR 86-5, Dalhousie University 1986

[32] J.H. DULÁ : An Upper Bound on the Expectation of Sublinear Functions of Multivariate Random Variables; Preprint, CORE (1987) - (forthcoming in Math. Progr.)

[33] J. DUPACOVÁ (-ZACKOVA) : On Minimax Solutions of Stochastic Linear Programming Problems ; Casopis pro Pestovani Mathematiky 91 (1966) 423-430

[34] J. DUPACOVÁ : Minimax Stochastic Programs with Nonconvex Separable Penalty Functions; in: A.Prékopa (ed.): Progress in Operations Research; North-Holland,Amsterdam (1976) 303-316

[35] J. DUPACOVÁ : Minimax Stochastic Programs with Non-Separable Penalties; in : K. Iracki , K. Malanowski, S. Walukiewicz (eds.) : Optimization Techniques, Part I ; Lecture Notes in Control and Information Science 22 (1980) 157-163

[36] J. DUPACOVÁ : The Minimax Approach to Stochastic Programming and an Illustrative Application; Stochastics 20 (1987) 149-162

[37] J. DUPACOVÁ : Stability and Sensitivity Analysis for Stochastic Programming; Annals of Operations Research 27 (1990) 115-142

[38] N.C.P. EDIRISINGHE and W.T. ZIEMBA : Bounds for Two-Stage Stochastic Programs with Fixed Recourse; Working Paper No. 89-MSC-024, Faculty of Commerce, University of British Columbia, Vancouver, Canada (submitted to Mathematics of Operations Research).

[39] I. EKELAND and R. TEMAM : Convex Analysis and Variational Problems; North-Holland 1976

[40] Y. ERMOLIEV : Stochastic Quasigradient Methods and their Applications to Systems Optimization; Stochastic 9 (1983) 1-36

[41] Y. ERMOLIEV : Stochastic Quasigradient Methods ; in: Y. Ermoliev and R.J.-B. Wets (eds.): Numerical Techniques for Stochastic Optimization; Springer-Verlag (1988) 141-186

[42] Y. ERMOLIEV and R.J.-B. WETS : Numerical Techniques for Stochastic Optimization; Springer-Verlag 1988

[43] Y. ERMOLIEV and R.J.-B. WETS : Stochastic Programming, an Introduction; in: Y. Ermoliev and R.J.-B. Wets (eds.): Numerical Techniques for Stochastic Optimization; Springer-Verlag (1988) 1-32

[44] A. FELLER: An Introduction to Probability Theory and its Application Vol. II; John Wiley & Sons 1966

[45] R. FLETCHER : Practical Methods of Optimization; John Wiley & Sons 1981

[46] K. FRAUENDORFER : Solving SLP Recourse Problems with Arbitrary Multivariate Distributions - The Dependent Case; Math. of OR 13/3 (1988) 377-394

[47] K. FRAUENDORFER : Solving SLP Recourse Problems: The Case of Stochastic Technology Matrix, RHS & Objective; in :M. Iri, K. Yajima (eds.) : Proceedings of 13th IFIP Conference on System Modelling and Optimization; Springer-Verlag (1988) 90-100

[48] K. FRAUENDORFER : SLP Problems : Objective and Right Hand Side being Stochastically Dependent - Part I; in : R. Vichnevetzky , P. Borne, J. Vignes (eds.): Proceedings of 12th IMACS World Congress in Paris, 2 (1988) 48-51

[49] K. FRAUENDORFER and P. KALL : A Solution Method for SLP Recourse Problems with Arbitrary Distributions - The Independent Case; Problems of Control and Information Theory 17 (1988) 177-205

[50] A. GAIVORONSKI : Implementation of Stochastic Quasigradient Methods; in: Y. Ermoliev and R.J.-B. Wets (eds.): Numerical Techniques for Stochastic Optimization; Springer-Verlag (1988) 313-351

[51] A. GAIVORONSKI : Interactive Program SQG-PC for Solving Stochastic Programming Problems on IBM PC/XT/AT/ Compatibles - User Guide; IIASA WP-88-11 , 1988

[52] H. I. GASSMANN : MSLiP: A Computer Code for the Multistage Stochastic Linear Programming Problem; Math. Progr. 47 (1990) 407-423

[53] H. GASSMANN and W.T. ZIEMBA : A Tight Upper Bound for the Expectation of a Convex Function of a Multivariate Random Variable; Math. Progr. Study 27 (1986) 39-53

[54] P.H.GILL, W. MURRAY and M.H.WRIGHT : Practical Optimization; Academic Press 1981

[55] P.R. HALMOS : Measure Theory; Springer-Verlag 1974

[56] J.L. HIGLE and S. SEN : Stochastic Decomposition: An Algorithm for Two-Stage Linear Programs with Recourse; Math. of OR 16/3 (1991) 650-669

[57] J.L. HIGLE and S. SEN : Statistical Verification of Optimality Conditions for Stochastic Programs with Recourse; in : J. Birge and R.J.-B. Wets (eds.) : Stochastic Programming, Part I; Annals of Operations Research 30 (1991) 215-240

[58] J.L. HIGLE and S. SEN: On the Convergence of Algorithms with Implications for Stochastic and Nondifferentiable Optimization; Working Paper, University of Arizona, 1989 (forthcoming in Mathematics of OR)

[59] J.L. HIGLE , S. SEN and D.S. YAKOWITZ : Finite Master Programs in Stochastic Decomposition; Working Paper, University of Arizona, 1990

[60] J.K. HO and E. LOUTE : A Set of Staircase Linear Programming Test Problems ; Math. Progr. 20 (1981) 91-94

[61] R. HORST and H. TUY : Global Optimization (Deterministic Approach); Springer-Verlag 1990

[62] C. HUANG, W.T. ZIEMBA and A. BEN-TAL : Bounds on the Expectation of a Convex Function of a Random Variable: with Application to Stochastic Programming; Operations Research 25 (1977) 315-325

[63] G. INFANGER : Monte Carlo (Importance) Sampling within Benders' Decomposition for Stochastic Linear Programs; Tech. Report SOL 89-13 (revised 1991) Departement of Operations Research, Stanford University CA - (forthcoming in Annals of Operations Research)

[64] J. JENSEN : Sur les fonctions convexes et les inégalités entre les valeurs moyennes; Acta Math. 30 (1906) 175-193

[65] P. KALL : Approximations to Stochastic Programs with Complete Fixed Recourse; Numer. Math. 22 (1974) 333-339.

[66] P. KALL : Stochastic Linear Programming; Springer-Verlag 1976

[67] P. KALL : Computational Methods for Solving Two-Stage Stochastic Linear Programming Problems; Zeitschrift für Angewandte Mathematik und Physik 30 (1979) 261-271

[68] P.KALL : Solving Complete Fixed Recourse Problems by Successive Discretization - Extended Abstract; in: P.Kall and A. Prékopa (eds.): Recent Results in Stochastic Programming; Lecture Notes in Economics and Mathematical Systems 179 (1980) 135-138

[69] P. KALL : Approximation to Optimization Problems: An Elementary Review; Math. of OR, 11/1 (1986) 9-17

[70] P. KALL : Stochastic Programs with Recourse: An Upper Bound and the Related Moment Problem; ZOR 31 (1987) A119-A141

[71] P. KALL : On Approximations and Stability in Stochastic Programming; in: J. Guddat, H. Th. Jongen, B. Kummer, F. Nožička (eds.) : Parametric Optimization and Related Topics; Akademie-Verlag (1987) 387-407

[72] P. KALL : Stochastic Programming with Recourse: Upper Bound and Moment Problems - A Review ; in : J. Guddat, B. Bank, H. Hollatz, P. Kall, D. Klatte, B. Kummer, K. Lommatzsch, K. Tammer, M. Vlach, K. Zimmermann (eds.): Advances in Mathematical Optimization; Akademie-Verlag (1988) 86-103

[73] P. KALL : An Upper Bound for SLP using First and Total Second Moments; in : J. Birge and R.J.-B. Wets (eds.) : Stochastic Programming, Part I; Annals of Operations Research 30 (1991) 267-276

[74] P. KALL and E. KELLER : GENSLP: A Program for Generating Input for Stochastic Linear Programs with Complete Fixed Recourse ; Manuscript, Institute for Operations Research University of Zurich 1983

[75] P. KALL and J. MAYER : A Model Management System for Stochastic Linear Programming - System Design; Manuscript, Institute for Operations Research (1991) - (forthcoming in: Proceedings of the 15th IFIP Conference on Modelling and Optimization, Zurich, 1991)

[76] P. KALL and W. OETTLI: Measurability Theorems for Stochastic Extremals; SIAM J. Control 13/5 (1975) 994-998

[77] P.KALL and A. PRÉKOPA : Recent Results in Stochastic Programming; Proceedings Oberwolfach 1979 ; Lecture Notes in Economics and Mathematical Systems 179, Springer-Verlag 1980

[78] P. KALL, A. RUSZCZYŃSKI and K. FRAUENDORFER : Approximation Techniques in Stochastic Programming; in: Y. Ermoliev and R.J.-B. Wets (eds.): Numerical Techniques for Stochastic Optimization; Springer-Verlag (1988) 33-64

[79] P. KALL and D. STOYAN : Solving Programming Problems with Recourse Including Error Bound; Mathematische Operationsforschung und Statistik, Ser. Opt. 13 (1982) 431-447

[80] J.L. KELLEY and I. NAMIOKA : Linear Topological Spaces; Springer-Verlag 1976

[81] A.J. KING : An Implementation of the Lagrangian Finite Generation Method; in: Y. Ermoliev and R.J.-B. Wets (eds.): Numerical Techniques for Stochastic Optimization; Springer-Verlag (1988) 295-312

[82] W. KLEIN HANEVELD : Duality in Stochastic Linear and Dynamic Programming; Lecture Notes in Economics and Mathematical Systems 274 , Springer Verlag 1986

[83] M. KREIN and A. NUDELMAN : The Markow Moment Problem and Extremal Problems; Translations of Mathematical Monographs 50 , American Mathematical Society, Providence 1977)

[84] F. LOUVAUX and Y. SMEERS : Optimal Investments for Electricity Generation: A Stochastic Model and a Test-Problem; in: Y. Ermoliev and R.J.-B. Wets (eds.): Numerical Techniques for Stochastic Optimization; Springer-Verlag (1988) 445-454

[85] A. MADANSKY : Bounds of the Expectation of a Convex Function of a Multivariate Random Variable; Annals of Mathematical Statistics 30 (1959) 743-746

[86] A. MADANSKY : Inequalities for Stochastic Linear Programming Problems; Management Science 6 (1960) 197-204

[87] R.E. MARSTEN : The Design of XMP Linear Programming Library; ACM Transactions on Mathematical Software 7/4 (1981) 481-497

[88] K. MARTI : Solving Stochastic Linear Programs by Semi-Stochastic Approximation Algorithms; in: P. Kall, A. Prékopa (eds.): Recent Results in Stochastic Programming; Lecture Notes in Economics and Mathematical Systems 179 (1980) 191-213

[89] K. MARTI : Computation of Descent Direction in Stochastic Optimization Problems with Invariant Distribution; ZAMM 65 (1985) 355-378

[90] K. MARTI : Descent Directions and Efficient Solutions in Discretely Distributed Stochastic Programs; Lecture Notes in Economics and Mathematical Systems 299, Springer -Verlag 1988

[91] K. MARTI and E. FUCHS : Rates of Convergence of Semi-Stochastic Approximation Procedures for Solving Stochastic Optimization Problems; Optimization 17 (1986) 243-265

[92] K. MARTI and E. FUCHS : Computation of Descent Directions and Efficient Points in Stochastic Optimization Problems Without Using Derivatives; Math. Progr. Study 28 (1986) 132-156

[93] K. MARTI and E. PLÖCHINGER : Optimal Stepsizes in Semi-Stochastic Approximation Procedures I; Optimization 21/1 (1990) 123-153

[94] K. MARTI and E. PLÖCHINGER : Optimal Stepsizes in Semi-Stochastic Approximation Procedures II; Optimization 21/2 (1990) 281-312

[95] L. MCLINDEN : Dual Operations on Saddle Functions; Transactions of the Mathematical Society 179 (1973) 363-381

[96] L. MCLINDEN : An Extension of Fenchel's Duality Theorem to Saddle Functions and Dual Minimax Problems; Pacific Journal of Mathematics 50/1 (1974) 135-158

[97] J.M. MULVEY and H. VLADIMIROU : Solving Multistage Stochastic Networks : An Application of Scenario Aggregation ; Report SOR-88-1, Dept. of Civil Engineering and Operations Research, Princeton University (July 1989) - (forthcoming in Networks)

[98] J.M. MULVEY and H. VLADIMIROU : Applying the Progressive Hedging Algorithm to Stochastic Generalized Networks; in : J. Birge and R.J.-B. Wets (eds.) : Stochastic Programming, Part II; Annals of Operations Research 31 (1991) 399-424

[99] B. MURTAGH and M. SAUNDERS : Large-Scale Linearly Constrained Optimization; Mathematical Programming 14 (1978) 41-72

[100] B. MURTAGH and M. SAUNDERS : MINOS 5.0 User's Guide; Report SOL 83-20, Department of Operations Research, Stanford University, California 1983 (Revised as MINOS 5.1 User's Guide, Report SOL 83-20R, 1987)

[101] L. NAZARETH : Design and Implementation of a Stochastic Programming Optimizer with Recourse and Tenders ; in: Y. Ermoliev and R.J.-B. Wets (eds.): Numerical Techniques for Stochastic Optimization; Springer-Verlag (1988) 273-294

[102] L. NAZARETH and R.J.-B. WETS : Algorithms for Stochastic Programs: The Case of Nonstochastic Tenders; Math. Progr. Study 28 (1986) 1-28

[103] L. NAZARETH and R.J.-B. WETS: Nonlinear Programming Techniques Applied to Stochastic Programs with Recourse; in: Y. Ermoliev and R.J.-B. Wets (eds.): Numerical Techniques for Stochastic Optimization; Springer-Verlag (1988) 95-122

[104] G.CH. PFLUG : Stepsize Rules, Stopping Times and Their Implementation in Stochastic Quasigradient Algorithms; in: Y. Ermoliev and R.J.-B. Wets (eds.): Numerical Techniques for Stochastic Optimization; Springer-Verlag (1988) 353-372

[105] A. PRÉKOPA : Contributions to the Theory of Stochastic Programming; Math. Prog. 4 (1973) 202-221

[106] A. PRÉKOPA : Programming under Probabilistic Constraints with a Random Technology Matrix; Math. Operationsforsch. Statist. 5 (1974) 109-116

[107] A. PRÉKOPA : Logarithmic Concave Measures and Related Topics; in : M. Dempster (ed.) Stochastic Programming; Academic Press (1980) 63-82

[108] A. PRÉKOPA : Numerical Solution of Probabilistic Constrained Programming Problems ; in: Y. Ermoliev and R.J.-B. Wets (eds.): Numerical Techniques for Stochastic Optimization; Springer-Verlag (1988) 123-139

[109] A. PRÉKOPA : The Discrete Moment Problem and Linear Programming ; Discr. Appl. Math. 27 (1990) 235-254

[110] A. PRÉKOPA and R.J.-B. WETS : Stochastic Programming 84:I&II; Math. Progr. Study 27-28, 1986

[111] S.M. ROBINSON : Local Epi-Continuity and Local Optimization; Math. Progr. 37 (1987) 208-222

[112] S.M. ROBINSON : Extended Scenario Analysis; in : J. Birge and R.J.-B. Wets (eds.) : Stochastic Programming, Part II; Annals of Operations Research 31 (1991) 385-398

[113] S.M. ROBINSON and R.J.-B. WETS : Stability in Two-Stage Stochastic Programming; SIAM J. Control and Optimization 25 (1987) 1409-1416

[114] R.T. ROCKAFELLAR : Minimax Theorems and Conjugate Saddle Functions; Math. Scand. 14 (1964) 151-173

[115] R.T. ROCKAFELLAR : Duality and Stability in Extremum Problems Involving Convex Functions; Pacific Journal of Math. 21/1 (1967) 167-187

[116] R.T. ROCKAFELLAR : Integrals which are Convex Functionals; Pacific Journal of Mathematics 24/3 (1968) 525-539

[117] R.T. ROCKAFELLAR : Measurable Dependence of Convex Sets and Functions on Parameters; Journal of Math. Anal. and Appl. 28 (1969) 4-25

[118] R. T. ROCKAFELLAR : Convex Analysis; Princeton University Press 1970

[119] R.T. ROCKAFELLAR : Conjugate Duality and Optimization; SIAM Monographs Series in Applied Mathematics, Philadelphia 1974

[120] R.T. ROCKAFELLAR : Augmented Lagrangians and Applications of the Proximal Point Algorithm in Convex Programming; Math. of OR 1 (1976) 97-116

[121] R.T. ROCKAFELLAR : Integral Functionals, Normal Integrand and Measurable Selections; in: Nonlinear Operators and the Calculus of Variations; Lecture Notes in Mathematics 543, Springer-Verlag (1976) 157-207

[122] R.T. ROCKAFELLAR : Computational Schemes for Large-Scale Problems in Extended Linear-Quadratic Programming ; Math. Progr. 48/3 (1990) 170-187

[123] R.T. ROCKAFELLAR : Large-Scale Extended Linear-Quadratic Programming and Multistage Optimization; in : S.Gomez, P. Hennert, R.A. Tapia (eds.): Advances in Numerical Partial Differential Equations and Optimization; SIAM Publications, Philadelphia (1991)

[124] R.T. ROCKAFELLAR and R.J.-B. WETS : Linear-Quadratic Programming Problems with Stochastic Penalties: The Finite Generation Algorithm; in: V.I.Arkin, A.Shiraer and R.J.-B. Wets (eds.): Stochastic Optimization; Lecture Notes in Control and Information Science, IIASA Series No 81, Springer-Verlag (1986) 545-560

[125] R.T. ROCKAFELLAR and R.J.-B. WETS : A Lagrangian Finite Generation Technique for Solving Linear-Quadratic Problems in Stochastic Programming; Math. Progr. Study 28 (1986) 63-93

[126] R.T. ROCKAFELLAR and R.J.-B WETS : A Dual Strategy for the Implementation of the Aggregation Principle in Decision Making under Uncertainty; Working Paper, November 1990

[127] R.T. ROCKAFELLAR and R.J.-B WETS : Scenarios and Policy Aggregation in Optimization under Uncertainty; Math. of OR 16 (1991) 119-147

[128] W. RÖMISCH and R. SCHULTZ : Distribution Sensitivity in Stochastic Programming; Math. Progr. 50 (1991) 197-226

[129] W. RÖMISCH and R. SCHULTZ : Stability Analysis for Stochastic Programs; Annals of Operations Research 30 (1991) 241-266

[130] W. RÖMISCH and A. WALKOLBINGER : Obtaining Convergence Rates for Approximations in Stochastic Programming; in J. Guddat, H. Th. Jongen, B. Kummer, F. Nozicka (eds.) : Parametric Optimization and Related Topics; Akademie (1987) 327-343

[131] W. RUDIN : Analysis; Physika-Verlag 1980

[132] A. RUSZCZYŃSKI : A Regularized Decomposition Method for Minimizing a Sum of Polyhedral Functions; Math. Progr. 35 (1986) 309-333

[133] A. RUSZCZYŃSKI : A Linearization Method for Nonsmooth Stochastic Programming; Math. of OR 12 (1987) 32-49

[134] A. RUSZCZYŃSKI and W. SYSKI : A Method of Aggregate Stochastic Subgradients with On-Line Stepsize Rules for Convex Stochastic Programming Problems; Math. Progr.Study 28 (1986) 113-131

[135] J.K. SENGUPTA : Stochastic Programming - Methods and Applications; North-Holland 1972

[136] D. STOYAN : Comparison Methods for Queues and Other Stochastic Models; Wiley & Sons (1983)

[137] B. STRAZICKY : Computational Experience with an Algorithm for Discrete Recourse Problems; in : Dempster M. (ed.): Stochastic Programming; Academic Press (1980) 263-274

[138] G. TINTNER : Stochastic Linear Programming with Applications to Agricultural Economics; in: H.A. Antosiewicz (ed.) : Proc. Second Symposium in Linear Programming; Washington (1955) 197-207

[139] H. TUY : Effect of the Subdivision Strategy on Convergence and Efficiency of Some Global Optimization Algorithms; Journal of Global Optimization 1 (1991) 23-36

[140] R. VAN SLYKE and R.J.-B. WETS : L-Shaped Linear Program with Application to Optimal Control and Stochastic Linear Programs; SIAM Journal on Appl. Math. 17 (1969) 638-663

[141] R.J.-B. WETS : Stochastic Programs with Fixed Recourse : The Equivalent Deterministic Program; SIAM Review 16 (1974) 309-339

[142] R.J.-B. WETS : Stochastic Programming: Solution Techniques and Approximation Schemes; in: A. Bachem, M. Grötschel, B. Korte (eds.): Mathematical Programming, The State of the Art, Springer-Verlag (1983) 566-603

[143] R.J.-B. WETS : Large Scale Linear Programming Techniques in Stochastic Programming; in: Y. Ermoliev and R.J.-B. Wets (eds.): Numerical Techniques for Stochastic Optimization; Springer-Verlag (1988) 65-94

[144] R.J.-B. WETS : Stochastic Programming; in: G.L. Nemhauser, A.H.G. Rinnoy Kan, M.J. Todd (eds.) : Handbook on Operations Research and Management Science, Vol 1: Optimization; North-Holland (1989) 573-629

[145] R.J.-B. WETS : The Aggregation Principle in Scenario Analysis and Stochastic Optimization; in: S. Wallace (ed.) : Algorithms and Model Formulations in Mathematical Programming; Springer-Verlag, NATO ASI Vol. 51 (1989) 91-113

Lecture Notes in Economics and Mathematical Systems

For information about Vols. 1–210
please contact your bookseller or Springer-Verlag

Vol. 211: P. van den Heuvel, The Stability of a Macroeconomic System with Quantity Constraints. VII, 169 pages. 1983.

Vol. 212: R. Sato and T. Nôno, Invariance Principles and the Structure of Technology. V, 94 pages. 1983.

Vol. 213: Aspiration Levels in Bargaining and Economic Decision Making. Proceedings, 1982. Edited by R. Tietz. VIII, 406 pages. 1983.

Vol. 214: M. Faber, H. Niemes und G. Stephan, Entropie, Umweltschutz und Rohstoffverbrauch. IX, 181 Seiten. 1983.

Vol. 215: Semi-Infinite Programming and Applications. Proceedings, 1981. Edited by A.V. Fiacco and K.O. Kortanek. XI, 322 pages. 1983.

Vol. 216: H.H. Müller, Fiscal Policies in a General Equilibrium Model with Persistent Unemployment. VI, 92 pages. 1983.

Vol. 217: Ch. Grootaert, The Relation Between Final Demand and Income Distribution. XIV, 105 pages. 1983.

Vol 218: P. van Loon, A Dynamic Theory of the Firm: Production, Finance and Investment. VII, 191 pages. 1983.

Vol. 219: E. van Damme, Refinements of the Nash Equilibrium Concept. VI. 151 pages. 1983.

Vol. 220: M. Aoki, Notes on Economic Time Series Analysis: System Theoretic Perspectives. IX, 249 pages. 1983.

Vol. 221: S. Nakamura, An Inter-Industry Translog Model of Prices and Technical Change for the West German Economy. XIV, 290 pages. 1984.

Vol. 222: P. Meier, Energy Systems Analysis for Developing Countries. VI, 344 pages. 1984.

Vol. 223: W. Trockel, Market Demand. VIII, 205 pages. 1984.

Vol. 224: M. Kiy, Ein disaggregiertes Prognosesystem fur die Bundesrepublik Deutschland. XVIII, 276 Seiten. 1984.

Vol. 225: T.R. von Ungern-Sternberg, Zur Analyse von Märkten mit unvollständiger Nachfragerinformaton. IX, 125 Seiten. 1984.

Vol. 226: Selected Topics in Operations Research and Mathematical Economics. Proceedings, 1963. Edited by G. Hammer and D. Pallaschke IX, 478 pages. 1984.

Vol. 227: Risk and Capital. Proceedings, 1983. Edited by G. Bamberg and K. Spremann VII, 306 pages. 1984.

Vol. 228: Nonlinear Models of Fluctuating Growth. Proceedings, 1983. Edited by R.M. Goodwin, M. Krüger and A. Vercelli. XVII, 277 pages. 1984.

Vol. 229: Interactive Decision Analysis. Proceedings, 1983. Edited by M. Grauer and A.P. Wierzbicki. VIII, 269 pages. 1984.

Vol. 230: Macro-Economic Planning with Conflicting Goals. Proceedings, 1982. Edited by M. Despontin, P. Nijkamp and J. Spronk. VI, 297 pages. 1984.

Vol. 231: G.F. Newell, The M/M/8 Service System with Ranked Servers in Heavy Traffic. XI, 126 pages. 1984.

Vol. 232: L. Bauwens, Bayesian Full Information Analysis of Simultaneous Equation Models Using Integration by Monte Carlo. VI, 114 pages. 1984.

Vol. 233: G. Wagenhals, The World Copper Market. XI, 190 pages. 1984.

Vol. 234: B.C. Eaves, A Course in Triangulations for Solving Equations with Deformations. III, 302 pages. 1984.

Vol. 235: Stochastic Models in Reliability Theory Proceedings, 1984. Edited by S. Osaki and Y. Hatoyama. VII, 212 pages. 1984.

Vol. 236: G. Gandolfo, P.C. Padoan, A Disequilibrium Model of Real and Financial Accumulation in an Open Economy. VI, 172 pages. 1984.

Vol. 237: Misspecification Analysis. Proceedings, 1983. Edited by T.K. Dijkstra. V, 129 pages. 1984.

Vol. 238: W. Domschke, A. Drexl, Location and Layout Planning. IV, 134 pages. 1985.

Vol. 239: Microeconomic Models of Housing Markets. Edited by K. Stahl. VII, 197 pages. 1985.

Vol. 240: Contributions to Operations Research. Proceedings, 1984. Edited by K. Neumann and D. Pallaschke. V, 190 pages. 1985.

Vol. 241: U. Wittmann, Das Konzept rationaler Preiserwartungen. XI, 310 Seiten. 1985.

Vol. 242: Decision Making with Multiple Objectives. Proceedings, 1984. Edited by Y.Y. Haimes and V. Chankong. XI, 571 pages. 1985.

Vol. 243: Integer Programming and Related Areas. A Classified Bibliography 1981–1984. Edited by R. von Randow. XX, 386 pages. 1985.

Vol. 244: Advances in Equilibrium Theory. Proceedings, 1984. Edited by C.D. Aliprantis, O. Burkinshaw and N.J. Rothman. II, 235 pages. 1985.

Vol. 245: J.E.M. Wilhelm, Arbitrage Theory. VII, 114 pages. 1985.

Vol. 246: P.W. Otter, Dynamic Feature Space Modelling, Filtering and Self-Tuning Control of Stochastic Systems. XIV, 177 pages.1985.

Vol. 247: Optimization and Discrete Choice in Urban Systems. Proceedings, 1983. Edited by B.G. Hutchinson, P. Nijkamp and M. Batty VI, 371 pages. 1985.

Vol. 248: Pural Rationality and Interactive Decision Processes. Proceedings, 1984. Edited by M. Grauer, M. Thompson and A.P. Wierzbicki. VI, 354 pages. 1985.

Vol. 249: Spatial Price Equilibrium: Advances in Theory, Computation and Application. Proceedings, 1984. Edited by P.T. Harker. VII, 277 pages. 1985.

Vol. 250: M. Roubens, Ph. Vincke, Preference Modelling. VIII, 94 pages. 1985.

Vol. 251: Input-Output Modeling. Proceedings, 1984. Edited by A. Smyshlyaev. VI, 261 pages. 1985.

Vol. 252: A. Birolini, On the Use of Stochastic Processes in Modeling Reliability Problems. VI, 105 pages. 1985.

Vol. 253: C. Withagen, Economic Theory and International Trade in Natural Exhaustible Resources. VI, 172 pages. 1985.

Vol. 254: S. Müller, Arbitrage Pricing of Contingent Claims. VIII, 151 pages. 1985.

Vol. 255: Nondifferentiable Optimization: Motivations and Applications. Proceedings, 1984. Edited by V.F. Demyanov and D. Pallaschke. VI, 350 pages. 1985.

Vol. 256: Convexity and Duality in Optimization. Proceedings, 1984. Edited by J. Ponstein. V, 142 pages. 1985.

Vol. 257: Dynamics of Macrosystems. Proceedings, 1984. Edited by J.-P. Aubin, D. Saari and K. Sigmund. VI, 280 pages. 1985.

Vol. 258: H. Funke, Eine allgemeine Theorie der Polypol- und Oligopolpreisbildung. III, 237 pages. 1985.

Vol. 259: Infinite Programming. Proceedings, 1984. Edited by E.J. Anderson and A.B. Philpott. XIV, 244 pages. 1985.

Vol. 260: H.-J. Kruse, Degeneracy Graphs and the Neighbourhood Problem. VIII, 128 pages. 1986.

Vol. 261: Th.R. Gulledge, Jr., N.K. Womer, The Economics of Made-to-Order Production. VI, 134 pages. 1986.

Vol. 262: H.U. Buhl, A Neo-Classical Theory of Distribution and Wealth. V, 146 pages. 1986.

Vol. 263: M. Schäfer, Resource Extraction and Market Struucture. XI, 154 pages. 1986.

Vol. 264: Models of Economic Dynamics. Proceedings, 1983. Edited by H.F. Sonnenschein. VII, 212 pages. 1986.

Vol. 265: Dynamic Games and Applications in Economics. Edited by T. Basar. IX, 288 pages. 1986.

Vol. 266: Multi-Stage Production Planning and Inventory Control. Edited by S. Axsäter, Ch. Schneeweiss and E. Silver. V, 264 pages.1986.

Vol. 267: R. Bemelmans, The Capacity Aspect of Inventories. IX, 165 pages. 1986.

Vol. 268: V. Firchau, Information Evaluation in Capital Markets. VII, 103 pages. 1986.

Vol. 269: A. Borglin, H. Keiding, Optimality in Infinite Horizon Economies. VI, 180 pages. 1986.

Vol. 270: Technological Change, Employment and Spatial Dynamics. Proceedings, 1985. Edited by P. Nijkamp. VII, 466 pages. 1986.

Vol. 271: C. Hildreth, The Cowles Commission in Chicago, 1939–1955. V, 176 pages. 1986.

Vol. 272: G. Clemenz, Credit Markets with Asymmetric Information. VIII,212 pages. 1986.

Vol. 273: Large-Scale Modelling and Interactive Decision Analysis. Proceedings, 1985. Edited by G. Fandel, M. Grauer, A. Kurzhanski and A.P. Wierzbicki. VII, 363 pages. 1986.

Vol. 274: W.K. Klein Haneveld, Duality in Stochastic Linear and Dynamic Programming. VII, 295 pages. 1986.

Vol. 275: Competition, Instability, and Nonlinear Cycles. Proceedings, 1985. Edited by W. Semmler. XII, 340 pages. 1986.

Vol. 276: M.R. Baye, D.A. Black, Consumer Behavior, Cost of Living Measures, and the Income Tax. VII, 119 pages. 1986.

Vol. 277: Studies in Austrian Capital Theory, Investment and Time. Edited by M. Faber. VI, 317 pages. 1986.

Vol. 278: W.E. Diewert, The Measurement of the Economic Benefits of Infrastructure Services. V, 202 pages. 1986.

Vol. 279: H.-J. Büttler, G. Frei and B. Schips, Estimation of Disequilibrium Modes. VI, 114 pages. 1986.

Vol. 280: H.T. Lau, Combinatorial Heuristic Algorithms with FORTRAN. VII, 126 pages. 1986.

Vol. 281: Ch.-L. Hwang, M.-J. Lin, Group Decision Making under Multiple Criteria. XI, 400 pages. 1987.

Vol. 282: K. Schittkowski, More Test Examples for Nonlinear Programming Codes. V, 261 pages. 1987.

Vol. 283: G. Gabisch, H.-W. Lorenz, Business Cycle Theory. VII, 229 pages. 1987.

Vol. 284: H. Lütkepohl, Forecasting Aggregated Vector ARMA Processes. X, 323 pages. 1987.

Vol. 285: Toward Interactive and Intelligent Decision Support Systems. Volume 1. Proceedings, 1986. Edited by Y. Sawaragi, K. Inoue and H. Nakayama. XII, 445 pages. 1987.

Vol. 286: Toward Interactive and Intelligent Decision Support Systems. Volume 2. Proceedings, 1986. Edited by Y. Sawaragi, K. Inoue and H. Nakayama. XII, 450 pages. 1987.

Vol. 287: Dynamical Systems. Proceedings, 1985. Edited by A.B. Kurzhanski and K. Sigmund. VI, 215 pages. 1987.

Vol. 288: G.D. Rudebusch, The Estimation of Macroeconomic Disequilibrium Models with Regime Classification Information. VII,128 pages. 1987.

Vol. 289: B.R. Meijboom, Planning in Decentralized Firms. X, 168 pages. 1987.

Vol. 290: D.A. Carlson, A. Haurie, Infinite Horizon Optimal Control. XI, 254 pages. 1987.

Vol. 291: N. Takahashi, Design of Adaptive Organizations. VI, 140 pages. 1987.

Vol. 292: I. Tchijov, L. Tomaszewicz (Eds.), Input-Output Modeling. Proceedings, 1985. VI, 195 pages. 1987.

Vol. 293: D. Batten, J. Casti, B. Johansson (Eds.), Economic Evolution and Structural Adjustment. Proceedings, 1985. VI, 382 pages.

Vol. 294: J. Jahn, W. Knabs (Eds.), Recent Advances and Historical Development of Vector Optimization. VII, 405 pages. 1987.

Vol. 295. H. Meister, The Purification Problem for Constrained Games with Incomplete Information. X, 127 pages. 1987.

Vol. 296: A. Börsch-Supan, Econometric Analysis of Discrete Choice. VIII, 211 pages. 1987.

Vol. 297: V. Fedorov, H. Läuter (Eds.), Model-Oriented Data Analysis. Proceedings, 1987. VI, 239 pages. 1988.

Vol. 298: S.H. Chew, Q. Zheng, Integral Global Optimization. VII, 179 pages. 1988.

Vol. 299: K. Marti, Descent Directions and Efficient Solutions in Discretely Distributed Stochastic Programs. XIV, 178 pages. 1988.

Vol. 300: U. Derigs, Programming in Networks and Graphs. XI, 315 pages. 1988.

Vol. 301: J. Kacprzyk, M. Roubens (Eds.), Non-Conventional Preference Relations in Decision Making. VII, 155 pages. 1988.

Vol. 302: H.A. Eiselt, G. Pederzoli (Eds.), Advances in Optimization and Control. Proceedings, 1986. VIII, 372 pages. 1988.

Vol. 303: F.X. Diebold, Empirical Modeling of Exchange Rate Dynamics. VII, 143 pages. 1988.

Vol. 304: A. Kurzhanski, K. Neumann, D. Pallaschke (Eds.), Optimization, Parallel Processing and Applications. Proceedings, 1987. VI, 292 pages. 1988.

Vol. 305: G.-J.C.Th. van Schijndel, Dynamic Firm and Investor Behaviour under Progressive Personal Taxation. X, 215 pages.1988.

Vol. 306: Ch. Klein, A Static Microeconomic Model of Pure Competition. VIII, 139 pages. 1988.

Vol. 307: T.K. Dijkstra (Ed.), On Model Uncertainty and its Statistical Implications. VII, 138 pages. 1988.

Vol. 308: J.R. Daduna, A. Wren (Eds.), Computer-Aided Transit Scheduling. VIII, 339 pages. 1988.

Vol. 309: G. Ricci, K. Velupillai (Eds.), Growth Cycles and Multisectoral Economics: the Goodwin Tradition. III, 126 pages. 1988.

Vol. 310: J. Kacprzyk, M. Fedrizzi (Eds.), Combining Fuzzy Imprecision with Probabilistic Uncertainty in Decision Making. IX, 399 pages. 1988.

Vol. 311: R. Färe, Fundamentals of Production Theory. IX, 163 pages. 1988.

Vol. 312: J. Krishnakumar, Estimation of Simultaneous Equation Models with Error Components Structure. X, 357 pages. 1988.

Vol. 313: W. Jammernegg, Sequential Binary Investment Decisions. VI, 156 pages. 1988.

Vol. 314: R. Tietz, W. Albers, R. Selten (Eds.), Bounded Rational Behavior in Experimental Games and Markets. VI, 368 pages. 1988.

Vol. 315: I. Orishimo, G.J.D. Hewings, P. Nijkamp (Eds), Information Technology: Social and Spatial Perspectives. Proceedings 1986. VI, 268 pages. 1988.

Vol. 316: R.L. Basmann, D.J. Slottje, K. Hayes, J.D. Johnson, D.J. Molina, The Generalized Fechner-Thurstone Direct Utility Function and Some of its Uses. VIII, 159 pages. 1988.

Vol. 317: L. Bianco, A. La Bella (Eds.), Freight Transport Planning and Logistics. Proceedings, 1987. X, 568 pages. 1988.

Vol. 318: T. Doup, Simplicial Algorithms on the Simplotope. VIII, 262 pages. 1988.

Vol. 319: D.T. Luc, Theory of Vector Optimization. VIII, 173 pages. 1989.

Vol. 320: D. van der Wijst, Financial Structure in Small Business. VII, 181 pages. 1989.

Vol. 321: M. Di Matteo, R.M. Goodwin, A. Vercelli (Eds.), Technological and Social Factors in Long Term Fluctuations. Proceedings. IX, 442 pages. 1989.

Vol. 322: T. Kollintzas (Ed.), The Rational Expectations Equilibrium Inventory Model. XI, 269 pages. 1989.

Vol. 323: M.B.M. de Koster, Capacity Oriented Analysis and Design of Production Systems. XII, 245 pages. 1989.

Vol. 324: I.M. Bomze, B.M. Pötscher, Game Theoretical Foundations of Evolutionary Stability. VI, 145 pages. 1989.

Vol. 325: P. Ferri, E. Greenberg, The Labor Market and Business Cycle Theories. X, 183 pages. 1989.

Vol. 326: Ch. Sauer, Alternative Theories of Output, Unemployment, and Inflation in Germany: 1960–1985. XIII, 206 pages. 1989.

Vol. 327: M. Tawada, Production Structure and International Trade. V, 132 pages. 1989.

Vol. 328: W. Güth, B. Kalkofen, Unique Solutions for Strategic Games. VII, 200 pages. 1989.

Vol. 329: G. Tillmann, Equity, Incentives, and Taxation. VI, 132 pages. 1989.

Vol. 330: P.M. Kort, Optimal Dynamic Investment Policies of a Value Maximizing Firm. VII, 185 pages. 1989.

Vol. 331: A. Lewandowski, A.P. Wierzbicki (Eds.), Aspiration Based Decision Support Systems. X, 400 pages. 1989.

Vol. 332: T.R. Gulledge, Jr., L.A. Litteral (Eds.), Cost Analysis Applications of Economics and Operations Research. Proceedings. VII, 422 pages. 1989.

Vol. 333: N. Dellaert, Production to Order. VII, 158 pages. 1989.

Vol. 334: H.-W. Lorenz, Nonlinear Dynamical Economics and Chaotic Motion. XI, 248 pages. 1989.

Vol. 335: A.G. Lockett, G. Islei (Eds.), Improving Decision Making in Organisations. Proceedings. IX, 606 pages. 1989.

Vol. 336: T. Puu, Nonlinear Economic Dynamics. VII, 119 pages. 1989.

Vol. 337: A. Lewandowski, I. Stanchev (Eds.), Methodology and Software for Interactive Decision Support. VIII, 309 pages. 1989.

Vol. 338: J.K. Ho, R.P. Sundarraj, DECOMP: an Implementation of Dantzig-Wolfe Decomposition for Linear Programming. VI, 206 pages.

Vol. 339: J. Terceiro Lomba, Estimation of Dynamic Econometric Models with Errors in Variables. VIII, 116 pages. 1990.

Vol. 340: T. Vasko, R. Ayres, L. Fontvieille (Eds.), Life Cycles and Long Waves. XIV, 293 pages. 1990.

Vol. 341: G.R. Uhlich, Descriptive Theories of Bargaining. IX, 165 pages. 1990.

Vol. 342: K. Okuguchi, F. Szidarovszky, The Theory of Oligopoly with Multi-Product Firms. V, 167 pages. 1990.

Vol. 343: C. Chiarella, The Elements of a Nonlinear Theory of Economic Dynamics. IX, 149 pages. 1990.

Vol. 344: K. Neumann, Stochastic Project Networks. XI, 237 pages. 1990.

Vol. 345: A. Cambini, E. Castagnoli, L. Martein, P Mazzoleni, S. Schaible (Eds.), Generalized Convexity and Fractional Programming with Economic Applications. Proceedings, 1988. VII, 361 pages. 1990.

Vol. 346: R. von Randow (Ed.), Integer Programming and Related Areas. A Classified Bibliography 1984–1987. XIII, 514 pages. 1990.

Vol. 347: D. Ríos Insua, Sensitivity Analysis in Multi-objective Decision Making. XI, 193 pages. 1990.

Vol. 348: H. Störmer, Binary Functions and their Applications. VIII, 151 pages. 1990.

Vol. 349: G.A. Pfann, Dynamic Modelling of Stochastic Demand for Manufacturing Employment. VI, 158 pages. 1990.

Vol. 350: W.-B. Zhang, Economic Dynamics. X, 232 pages. 1990.

Vol. 351: A. Lewandowski, V. Volkovich (Eds.), Multiobjective Problems of Mathematical Programming. Proceedings, 1988. VII, 315 pages. 1991.

Vol. 352: O. van Hilten, Optimal Firm Behaviour in the Context of Technological Progress and a Business Cycle. XII, 229 pages. 1991.

Vol. 353: G. Ricci (Ed.), Decision Processes in Economics. Proceedings, 1989. III, 209 pages 1991.

Vol. 354: M. Ivaldi, A Structural Analysis of Expectation Formation. XII, 230 pages. 1991.

Vol. 355: M. Salomon. Deterministic Lotsizing Models for Production Planning. VII, 158 pages. 1991.

Vol. 356: P. Korhonen, A. Lewandowski, J . Wallenius (Eds.), Multiple Criteria Decision Support. Proceedings, 1989. XII, 393 pages. 1991.

Vol. 357: P. Zörnig, Degeneracy Graphs and Simplex Cycling. XV, 194 pages. 1991.

Vol. 358: P. Knottnerus, Linear Models with Correlated Disturbances. VIII, 196 pages. 1991.

GPSR Compliance

The European Union's (EU) General Product Safety Regulation (GPSR) is a set of rules that requires consumer products to be safe and our obligations to ensure this.

If you have any concerns about our products, you can contact us on ProductSafety@springernature.com

In case Publisher is established outside the EU, the EU authorized representative is:

Springer Nature Customer Service Center GmbH
Europaplatz 3
69115 Heidelberg, Germany

Batch number: 10844829

Printed by Printforce, the Netherlands